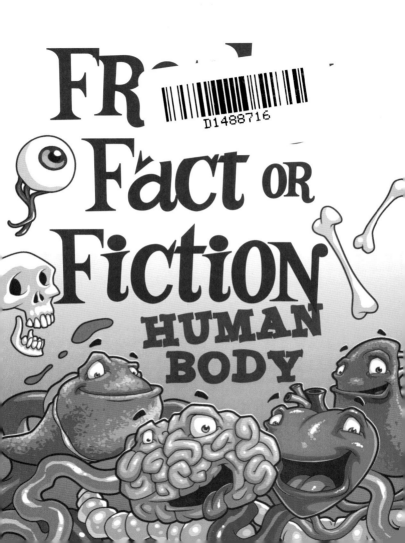

Published by Hinkler Books Pty Ltd
45–55 Fairchild Street
Heatherton Victoria 3202 Australia
www.hinkler.com.au

© Hinkler Books Pty Ltd 2011

Author: Claire Saxby
Editor: Suzannah Pearce
Copyeditors: Helena Newton and Susie Ashworth
Design: Diana Vlad and Ruth Comey
Cover Illustration: Rob Kiely
Illustrations: Brijbasi Art Press Ltd
Typesetting: MPS Limited
Prepress: Splitting Image

The publisher has made every effort to ensure that the facts and figures in this
book are correct at the time of publication. The publisher is not responsible for the
content, information, images or services that may appear in any books, journals,
newspapers, websites or links referenced.

ISBN: 978 1 7418 5255 4

Printed and bound in China

Freaky Fact or Fiction

They say that truth is stranger than fiction . . . but can you tell the difference? This book contains over 200 strange and interesting stories about the human body. Most of these are true, but some are tall tales; it will take an expert to spot the difference.

Quiz yourself, your parents, your little sister or your best friend. You can record your answers by ticking one of the circles at the bottom of each page. Then, to check whether you were right, turn to the answers section at the end of the 'facts'.

For extra fun, we've included our sources at the very end of this book. If you want to read more about the amazing human body, or if you just want to double-check a fact that sounds crazy, the sources are a good place to begin your research.

You can start anywhere in the book and read the facts in any order. Whatever you do, get ready for hours of *Freaky Fact or Fiction* fun!

Freaky Fact or Fiction

1 Adult skeletons have 206 bones, but babies are born with about 300. You are born with your bones only partly formed, and some of them are invisible on X-ray. As you grow, your bones lengthen and change shape. As you grow up, some bones also fuse together with others to become one. For example, in the thigh, five bones fuse to become the femur. Some bones will keep growing until you are about 25 years old.

 ✓ FACT OR FICTION

2 You put food into your mouth (ideally a little at a time), chew and then swallow the bits. The stomach produces acid to break up the food into even smaller bits. This stomach acid is strong. Incredibly strong. In fact, it's so strong that your body has to regularly add a new layer to the surface of the stomach. If your body didn't do this, the acid would start to burn holes in the stomach lining, and would eventually burn right through the stomach wall. Ouch!

✓ **FACT** **OR** **FICTION**

Freaky Fact or Fiction

3

There are some strange measurement matches related to your body. Measurements that just don't sound right at all. Measure the distance between the crease on the inside of your elbow and the crease inside your wrist. Now measure your foot. If you're flexible enough, double-check and measure your foot against your forearm. It doesn't seem possible, but the length is the same.

✓ **FACT**   **OR**   **FICTION**

6

Many scientists say that humans and apes descended from a common ancestor – and like apes, our ancestor had opposable big toes as well as opposable thumbs. This meant the tip of the big toe could reach and touch the tip of all the other toes, just like you can do with your thumbs. This handy big toe made climbing and balancing easier. As humans evolved, the opposable toe gradually moved. These days all your toes are in line. But the muscles that moved the big toe sideways are still there.

✓ **FACT** **OR** **FICTION**

Freaky Fact or Fiction

7 Holding a brand-new baby is a tricky thing. They're little and squirmy. Although they have lots of muscles, they haven't learnt how to control them yet. They can't even hold up their head! Their arms and legs move, but they can do very little with them. Human muscle control develops from the head down. First, neck muscles strengthen so the baby can hold up its head and turn it from side-to-side. Then the baby learns to hold toys, sit up, crawl, stand and walk. Then run!

 ✓ FACT OR FICTION

8 After you die, everything stops. Your heart stops pumping and your blood stops circulating. Your nerves stop sending messages to your brain and back again. Your muscles stop twitching and your body will never need food again. Your skin dries out. But the energy left in your body makes your fingernails and toenails grow, and your hair grows longer, too.

✓ **FACT** **OR** **FICTION**

9

What's your favourite food? You can only taste its deliciousness because of tiny cells clustered all over your tongue. These tastebuds are fully developed on your tongue from the instant of your birth. Different tastebuds will detect different tastes. The five recognised groups of tastes are sweet, salty, sour, bitter and umami (savoury). Your body constantly replaces tastebuds, so you will always have more than 10 million on your tongue.

✓ **FACT**   **OR**   **FICTION**

10

Your lungs are under your ribs, above your stomach, below your throat. They take up most of the space in your chest. You have a pair of them: one lung on the right side of your body; one lung on the left. They are the same size and shape and are joined at the trachea (trak-eea), which is also called the windpipe. Each lung has three sections: the superior (upper) lobe, the middle lobe and the inferior (lower) lobe.

✓ **FACT** **OR** **FICTION**

11

Your brain is a seething mass of nerve cells. If you were looking at the brain under a very powerful microscope, it would look a bit like a super-full pit of snakes all tangled together. Unlike the snakes, each nerve knows exactly what it is doing and where it's going. Each nerve cell receives information at one end and passes it on at the other, like a giant network of Chinese Whispers or Telephone. Unlike this game, the message remains the same.

 ✓ FACT   OR   FICTION

12

The largest part of the brain is the cerebrum, which has two halves (called hemispheres) and sits just beneath the skull. The cerebrum controls your body's movements and stores memories. It's the part of the brain that you see most often in pictures. It's roundish, greyish and full of folds. Have you ever seen a walnut in its shell? Your brain looks just like that. The outer layer of the cerebrum contains 14,000 nerve cells. That's a lot of thinking!

 ✓ **FACT** **OR** **FICTION**

13

Information zooms around the body to and from the brain like a train speeding between stations. Think about how long it takes you to move your hand if it gets too close to something scorching hot. Or how long it takes to lift your foot if you stand on a prickle. Not long. In that time the message has gone from your foot to your brain and back again. At its slowest, the message moves at about 3.6 km/h (2.25 mi/h). At its fastest, the message zooms along at around 108 km/h (67.5 mi/h), about the same speed as a car on a freeway.

 ✓ FACT   OR FICTION

An Adam's apple is called that because it looks like a small apple that has stuck in your throat. The Adam's apple is part of your larynx, also known as your voice box. If you are a boy, your Adam's apple will appear at the front of your neck during puberty, when every part of your body seems to be growing.

The cartilage of the Adam's apple also grows during puberty,

FICTION

Freaky Fact or Fiction

15 There are two tubes in your throat. One, the trachea (trak-eea), is for breathing and leads to your lungs. The other, the oesophagus (ee-sof-a-gus), is for food and leads to your stomach. It's not often that the wrong stuff goes into the wrong tube. Why? Because the food tube has a lid called the epiglottis (eppy-gloh-tus). It's a bit like a flip-top bin. The epiglottis closes over the food tube every time you swallow.

✓ **FACT**　　　　**OR**

16 Have you ever wondered what the fuss is about with Brussels sprouts? Why lots of kids hate them but some older people like them as much as chocolate? It's to do (partly) with tastebuds. Kids just have more — tastebuds that is. Each tastebud lives for about a week, then is replaced with a new one. In adults, some of the tastebuds are not replaced as quickly and sometimes not at all. So, older people have fewer tastebuds. Perhaps that explains Brussels sprouts. Or not.

✓ **FACT** **OR** **FICTION**

Freaky Fact or Fiction

17

real smile (rather than a pretend I'm-not-really-happy-but-I-know-you-want-me-to smile type) doesn't just involve your mouth but the whole of your face. Your mouth stretches, your cheeks are raised and you get little wrinkles (or crow's feet) in the corners of your eyes. If you want to be sure that someone is REALLY smiling, check to see if these other parts of their face move. If not, perhaps their heart isn't in it.

✓ FACT   OR   FICTION

18

Your brain grows faster than your face. Your brain and skull are almost completely grown by the time you are six years old. When you are born, four-fifths of your head is brain and skull and only one-fifth of your head is face. By the time you reach adulthood, your face is about half the size of your head.

✓ **FACT** **OR** **FICTION**

19

When you were a baby, you breathed faster than you do now, and much faster than you will when you are an adult. You breathed about 33 times a minute when you were a baby. By the time you are an adult your breathing will have slowed to less than half that rate. An adult breathes about 14 times a minute. For each baby-breathing minute, you breathed in about 500 mL (about 1 pt) of air. As an adult you will breathe in about 7 L (14 pt) every minute.

 FACT **OR** **FICTION**

20

abies are born without bony kneecaps. The kneecap is a special kind of bone called a sesamoid. It should be called the secret bone. Sesamoid bones form inside tendons that move over bony surfaces (like when your knee bends). The kneecap, or patella, develops inside the tendon at the front of your knee from when you are about two years old. It helps to make sure you can run and jump and skip without getting all wobbly.

✓ **FACT** OR **FICTION**

21

You have between 200,000 and 500,000 sweat glands on your skin. That means an average of between 15 and 34 per square centimetre (about 6 to 14 per square inch) of your skin. There are more on the palms of your hands and soles of your feet than other parts of your body. Mostly, sweat glands are there to cool your skin, but those on the palms and soles are designed to keep the skin moist and soft so you can hang on tight.

✓ **FACT**   **OR**   **FICTION**

22

There are about 150,000 hair follicles on the head. Each hair goes through three phases in its life. The first is an active growing phase, when the hair gets longer and longer. This phase can last for several years. The second is a resting phase, when the hair no longer grows. Perhaps it needs a snooze after all that growing. After the resting phase is finished, a new hair begins to grow, pushing out the old one. This is the third phase: shedding. Every day you shed about 50 to 100 hairs.

 FACT **OR** **FICTION**

23 A cough is a good thing. Really. A cough is a rapid explosion of air out of your lungs. Your body uses a cough reflex to make sure nothing but air gets into your lungs. If you accidentally breathe in something else, like dust, or a fly, or the mucus that comes with a cold, you will cough to push it straight back out again. Coughs have been measured at over 100 km/h (60 mi/h).

 FACT **OR**   **FICTION**

24 Most muscles in the body have a single job. For example, one muscle's job is to bend the elbow. Another muscle will straighten it. But the tongue is a special collection of muscle cells with many jobs to do. It moves food around the mouth while you are chewing. It ensures saliva is mixed with the food bits to make them moist. When chewing is finished, the tongue pushes the food to the back of your throat and into your oesophagus (ee-sof-a-gus) for swallowing. The tongue also helps you to talk.

✓ FACT OR FICTION

25

The outer ear, the part you can see and feel, is called the auricle. Around the auricle are several small muscles that attach it to the skull and scalp. These muscles are similar in all animals with outer ears. If you try hard, you can move your ears up and down and away from your head, just like a dog or a cat can.

 ✓ **FACT**   **OR**   **FICTION**

26

Y our ears and nose will continue to grow all your life. When you are born, your ears and nose are quite tiny. As the rest of you grows, they grow along with you. But most of you stops growing when you reach adulthood. Not your ears, or your nose. They will keep growing until the day you die.

 ✓ **FACT** **OR** **FICTION**

Freaky Fact or Fiction

You may be familiar with the names for many parts of your body. A foot is a foot. A nose is a nose. But there are names for parts of the body that you may not know as well. The name for the space between your eyebrows is called the glabella (glah-bell-a) and the name for your nasal passage is called the meatus (mee-ar-tus or mee-ate-us).

✓ FACT   OR   FICTION

28 Look at your baby photos. Look at your fingers, your toes, your nose. Most bits of you are tiny, tiny, tiny. Your body has lots of growing to do after you are born. But not your eyes. Your eyes are almost as big as they will ever be. Newborn babies' eyes are over 2/3 of the size that they will become in adulthood, so they really don't grow much more at all.

✓ **FACT** **OR** **FICTION**

29

There are 22 bones in your head. Eight form part of your cranium (the bit covered by hair) and 14 are part of your face. Together, the 22 bones are known as your skull. When you are born your cranium bones aren't completely formed and there are six soft spots on your head called fontanelles. The largest of these is on top of your head, towards the front. These soft spots disappear as your cranium bones join together and harden. They usually have all disappeared by the time you are three months old.

 ✓ FACT   OR   FICTION

30 Imagine if you grew fingernails on your head, or if hair sprouted from your fingertips. It would be so uncomfortable to sleep and hair would always be getting in your food. Luckily, it doesn't work that way. But fingernails, toenails and hair are made from the same cells, just arranged differently. Your fingernails grow twice as fast as your toenails do, and slower than your hair. Fingernails take about three months to grow from the skin fold at the base to the free edge. Toenails take about twice that time. Hair grows about three times faster than your fingernails.

✓ **FACT** **OR** **FICTION**

31

Not all hair is the same. Not even on your body. Imagine if all the hairs on your body grew to the same length as your head hair. Hairdressers would love it! They'd be busy seven days a week, trimming your arm hairs and your leg hairs, as well as the hair on your head. And everyone, male and female, would have hair all over their face, even on their forehead. Your eyelashes would cover your eyes! But luckily, the body is cleverer than that. Eyelashes only grow long enough to keep dust out of your eyes.

 FACT **OR** **FICTION**

32

ook closely at a hair from your head. Compare it with a friend's hair. The colour might be different and you may be able to see a difference in thickness, but mostly it looks the same. Look under a microscope and you can see that they're not the same at all. If you cut across a hair, and look end-on, straight hair is flat. Wavy hair is round in some parts, oval in others. And very curly hair looks almost perfectly round.

✓ **FACT** **OR** **FICTION**

33 It's never a lot of fun to have a cold. But just for a minute, think about the tiny virus that causes the cold. Your body's first response to the virus is to create more and more mucus in your nose to try to flood out the virus. When that doesn't work, the irritation in the nose causes you to sneeze to try to explode the pesky thing out. Then you start coughing to eject the virus from your lungs. How unwanted must the virus feel?

✓ **FACT**   **OR**   **FICTION**

34

Vomiting is the forcible ejection of the stomach contents via your mouth. It is never much fun. Usually you feel rotten beforehand and sometimes, even afterwards. The feeling starts deep in your stomach and moves upwards until there's nowhere to go but out. Along the way, muscles squeeze, one after another to keep the vomit moving along, just like squeezing the stuffing out of a raw sausage. But faster.

✓ **FACT** **OR** **FICTION**

35

How is growing a beard like joining the dots? Both start with an outline and join up later. In puberty, boys begin to grow new hair all over their bodies, including on their faces. Facial hair appears in a particular order. The first bits appear above the corners of the mouth, then gradually above the top lip. Next the cheeks, then under the middle of the lower lip. Finally all the bits join up. Voila! A beard.

✓ FACT   OR   FICTION

36 Do you always eat everything on your plate? Even if you do, not everything you eat can be used by your body. The food moves through your stomach, small intestine and then your large intestine. In your large intestine (or bowel), bacteria helps to break down the last bit of usable food. Each day when you open your bowels (do number twos, or poo) you release more than three teaspoons full of bacteria.

✓ **FACT** **OR** **FICTION**

Freaky Fact or Fiction

37 Gurgle, gurgle, burble. Listen to your own belly, or to the belly of someone else. (It's easier to get close to someone else's belly!) That gurgling noise has a special name: borborygmi (bor-boh-rig-mee). It's gas moving through the intestine. It's part nitrogen, part oxygen, part hydrogen, part methane and part carbon dioxide. What a mix! Guess what the gas is called when it makes it all the way through the intestines and leaves your body? That's right – flatus (flay-tuss)!

✓ FACT   OR   FICTION

38 **M**ilk contains lactose, which is milk sugar. Not everyone can digest lactose. Those who can't break down lactose into usable bits suffer from lactose intolerance. If a lactose-intolerant person drinks a glass of milk, the milk travels all the way through to the large intestine without being digested. There it mixes with bacteria and makes gas. Lots of it. One cup of hydrogen gas for each glass of milk! Oh, poor stretched tummy. Only one way out for that gas.

 ✓ **FACT** **OR** **FICTION**

39

The small intestine is like a rubbery, slippery, twisty tube a bit fatter than a garden hose. It connects the stomach with the large intestine. If it were stretched out in one straight line, it would measure about 7 m (23 ft). But it's not stretched out. It's curled up inside your belly like one long loopy piece of spaghetti. That's longer than a skipping rope and about the same as 23 bits of spaghetti laid end to end.

✓ **FACT**   **OR**   **FICTION**

40

Do you get tired walking to school? Or walking home? Just remember to count the steps. If you work out how far each step is, you can work out how many kilometres (miles) you walk each day. If you add up the footsteps of a lifetime and convert them into kilometres (miles), you can work out how far you will walk in a lifetime. The average person will walk more than three times around the world. That's about 128,000 km (80,000 mi). Hope you're wearing comfortable shoes!

 ✓ **FACT** **OR**   **FICTION**

Freaky Fact or Fiction

41

Some people say that you can predict how tall someone will be by doubling the height they are at two years old. But one man in Austria, Adam Rainer (1899–1950), grew at his own rate. At 21 years old, Adam was only 1.18 m (3 ft 10.5 in) tall. Then he had a growth spurt. Over the next 10 years, he almost doubled his height, reaching 2.18 m (7 ft 1.75 in). If the prediction was true for him, Adam would have been about the same height at two years old that he was on his 21st birthday.

 ✓ **FACT** **OR**   **FICTION**

42

A tongue is a handy thing. It's great for licking an ice-cream, swallowing and talking. Not all tongues are the same. Just as there are tall people and tiny people, tongues are different sizes. The longest tongue ever measured was 9.8 cm (3.9 in) from the tip to the centre of the top lip. The tongue's owner was Stephen Taylor from the United Kingdom, who stuck his tongue out for posterity in February 2009. That's one man you wouldn't want to share your ice-cream with – he'd have it all in just one lick.

 FACT   **OR**   **FICTION**

43

How did you learn to count? Did you use your fingers and toes? Everyone has 10 fingers and 10 toes, right? Not quite. Most people do, but a few people have more, and a few have less. People with polydactyly (poll-ee-dac-til-ee) have more. The most digits ever recorded on a living person is 25. This record is shared by Pranamya Menaria (born 10 August 2005) and Devendra Harne (born 9 January 1995), both from India. The Wadoma and Kalanga tribes of Africa could use some of the extras. Some of them are born with only two toes on each foot.

 ✓ **FACT**   **OR**   **FICTION**

44

Have you got a very loud brother or sister? Or are you the loud one in the family? It's hard to be quiet all the time, especially when you have friends over to play, or for dinner. But who do you think could make the loudest burp? Male or female? You'd be surprised. It was a woman! The loudest burp ever recorded was very loud – nearly 105 decibels. That's almost as loud as a jackhammer or a power saw.

✓ **FACT** **OR** **FICTION**

45

Sharks keep growing new teeth all their lives but it doesn't work that way for humans. Baby teeth grow and fall out and permanent teeth arrive. Except for wisdom teeth, most of your permanent teeth will have arrived by the time you're a teenager. Then that's it. Once you're an adult, you're finished with growing teeth. Look after them because they're the only teeth you'll ever get. Single teeth can appear later, but it's rare. It happened to Mária Pozderka of Hungary (born 19 July 1938) – she got a new tooth when she was 68 years old.

✓ **FACT**   **OR**   **FICTION**

46 Imagine someone said to you, 'Here – you can have one handful of chocolates.' That's one of the times it would be great to have the biggest hands in the world. The biggest hands in the world belonged to American Robert Wadlow (1918–1940) and measured 32.4 cm (12.75 in) from the wrist to the tip of the middle finger. Even better would be if someone said you could have one shoeful of chocolates and you had the biggest feet in the world. Robert's feet measured 47 cm (18.5 in) in length.

 ✓ **FACT** **OR** **FICTION**

47

I scream, you scream, we all scream for ice-cream! Anyone serving you, me or us at the ice-cream counter might want to wear earphones. Or they might send us away without anything. Or tell us to ask again, but quietly. There's nothing gentle or quiet about a scream. The loudest scream recorded was by Jill Drake (UK) in 2000. At 129 dB, her scream was almost as loud as a shotgun firing, or a jet plane taking off. Ouch!

 ✓ FACT OR FICTION

48 Have you ever wanted to grow your hair? Cut out all that wasted time sitting in the hairdresser's chair? You'd better get started if you're hoping to set a new world record. The record for the longest hair is currently held by Xie Qiuping, a Chinese woman who has been growing her locks since she was 13 years old. When it was last measured in 2004 she was 44 years old and her hair was 5.627 m (18 ft 5.54 in) long.

 ✓ **FACT** **OR** **FICTION**

Freaky Fact or Fiction

49

There's nothing more irritating than an itch you just can't reach. Like an itch in the middle of your back. You can almost reach it over your shoulder but just not quite. Longer arms might help. Or longer fingernails. But you always have to cut your fingernails to keep them clean and tidy. Lee Redmond, a woman in the US, could scratch any itch she liked. In 2008, her fingernails were measured at a combined length of 8.65 m (28 ft 4.5 in). On average, that means each fingernail was around 86.5 cm (2 ft 10 in).

 ✓ **FACT** **OR** **FICTION**

50

The human body needs to be cleaned from time to time, and there's nothing like a hot, steaming bath at the end of a long, dirty day. Especially if it is a big bath and includes lots of bubbles. Perfect for a long soak. But there are some baths you might think twice (or more) about before hopping in. There are those who think a bath of ice is wonderful, and there are legends about baths in milk. But the worst bath would have to be the one shared with 87 snakes. Hard to imagine that being fun at all.

✓ **FACT** **OR** **FICTION**

Freaky Fact or Fiction

51

Seven litres of water would more than half-fill a bucket. It's more than enough to wash the breakfast dishes. It would fill a glass about 30 times, or 30 glasses once. It is enough water to handwash a car, and more than enough to quench a thirst. It's also the amount of digestive fluid released into the human digestive tract every day.

✓ FACT   OR   FICTION

54

idneys are the same shape as a kidney bean. About the same colour too, but bigger. Kidneys generally come in pairs, one sitting either side of your body close to your back. They are about 10 cm (4 in) long and filter (clean) all your blood (about 5 L or 1.3 gal) every 45 minutes. That adds up to more than 160 L (over 35 gal) of blood filtered each day. All that's left after the clean blood returns to the blood vessels is about 1.4 L (3 pt) of liquid that the body gets rid of as urine (wee or pee).

✓ **FACT** **OR** **FICTION**

Freaky Fact or Fiction

55

Does your body always do what you tell it to? Do you sometimes miss when you're trying to hit a cricket ball or tennis ball? Sometimes it seems like someone else is in charge. Your body has different types of muscles. Some are called 'voluntary' muscles, which means you can move them where you want. Other muscles, like the ones that keep your heart beating, and the ones that push your food through your stomach and intestine, are called 'involuntary'. They keep on working when you are awake or asleep. They keep on working even when you hit or miss that ball.

 ✓ FACT OR FICTION

56 In movies sometimes blood is shown spurting from a wound, bright and pulsing. This will only happen if the bleeding is from an artery, because only arteries are made with tiny muscles. Muscles squeeze the blood through the arteries and around the body and can be felt as a pulse. Veins have no muscles at all. If a vein is cut, the blood trickles out.

✓ FACT OR FICTION

Freaky Fact or Fiction

57

Doctors and nurses look for pulses as a way to see how the heart is going (or IF it is still going). All arteries will pulse, but they're not all easy to find. It's a bit tricky to find your own pulse because the fingers you use to feel for it will have their own faint pulse. Common places to feel the pulse are inside the wrist, at the side-front of the neck, at the temple and behind the knee.

✓ FACT   OR   FICTION

58 lood is red, right? Right. What about the blue veins that you can see through your skin? It isn't really blue. Blood in the veins is a dark red colour. It just shows blue through your pink skin. Blood in arteries is bright red. It's the amount of oxygen in the blood that changes the colour. The more oxygen, the brighter red the blood.

✓ **FACT** **OR** **FICTION**

Freaky Fact or Fiction

59

Lurking deep in your bloodstream are white blood cells called 'natural killers' or NK cells. Some white blood cells exist to kill any invaders such as viruses or infections – cells that don't belong in your blood. But natural killers are different. Their role is to turn on your body's own cells, then kill them. But don't worry, they only kill your own cells if the cells are damaged or diseased.

 ✓ FACT OR FICTION

60

There's nothing worse than having bugs crawling on your skin. Oh, yes there is! It's much worse having bugs crawling inside your body, making you sick. In the disease called schistosomiasis (shiss-to-so-my-uh-sis), a tiny flatworm squirms through your skin and lays between 30 and 50 eggs per day into the bloodstream. Your body evicts the eggs when you go to the toilet. Yay! The eggs hatch and grow (in snails) and then when they are big enough, the new flatworms can burrow back into your body to feed. Not so yay!

 FACT **OR** **FICTION**

61

What do humans and sheep have in common? Not their coats – sheep win there. Not their feet – sheep have twice as many feet and no hands. Not their conversation – sheep don't have a lot to say. Horns? Surprisingly that's what they have in common with humans. If you permanently damage a fingernail or a toenail, it will grow thick, ridgy and curved – just like a ram's horns do. In fact, toenails and fingernails that grow like that are called 'ram's horn nails'.

✓ **FACT** **OR** **FICTION**

62 If you thought hives were just for bees, think again. Hives are a very, very itchy, lumpy, reddish, whitish skin reaction to something. There's no honey in these hives! Sometimes the reaction is to a food, like peanuts or shellfish. But another form of hives is dermatographism (der-mat-o-graf-ism), or skin writing. When you 'write' on the sufferer's skin by touching it or stroking it with an object, the skin reacts by swelling and becoming red. You can then actually read their skin. It's a bit like a temporary tattoo.

 ✓ **FACT** **OR** **FICTION**

Freaky Fact or Fiction

63 Are you scared of spiders? Snakes? Going outside your home? When a fear begins to affect the way you live your life, it's called a phobia. Most phobias have names. There are well-known phobias like arachnophobia (fear of spiders) and claustrophobia (fear of enclosed spaces). But there are many others. How about tremophobia? That's a fear of trembling. People with limnophobia are scared of lakes. There's even a name for people who are scared of developing a phobia. They have phobophobia!

✓ FACT   OR   FICTION

64

Puberty (when your body begins to change from a child to an adult) is a bit like spring. Very difficult to predict and to measure. Hormone levels go up and down like the temperature. Hair appears like new leaves. All sorts of other bits grow like crazy. Parents complain about having to keep buying new shoes. Not everything keeps up though. Usually when you stand up, your body adjusts your blood flow so you don't get light-headed or dizzy. But if you grow taller too fast, sometimes if you stand up suddenly, you might feel a little dizzy, or you might faint.

✓ **FACT** **OR** **FICTION**

65

Plant some seeds in good soil, water them and they will grow. Hang on – there's one more thing you need. Sunlight. Without sunlight, or with not enough sunlight, plants will not grow strong, if at all. Luckily you are not a plant. As long as you have enough good food, water and exercise, you will grow strong and healthy.

✓ **FACT** **OR** **FICTION**

66

Sailing, even just for a few hours, can make you very hungry. It's all that fresh air and work moving the sails to the right position. But it's usually just for fun and you're never very far from shore. Or food. It was much more difficult for long-distance sailors in the days before there were fridges. They might have been at sea for months, and fresh fruit and vegetables were the first foods to run out. Some sailors developed scurvy from the lack of Vitamin C. They became very grumpy, their teeth fell out and they had no energy for anything.

 ✓ FACT **OR** **FICTION**

Freaky Fact or Fiction

67 **D**on't cry. It makes your eyes all red and puffy and your face all blotchy. Amazing, you might think, since tears are simply salty water. Actually they're not. Well, not completely. There is salty water in tears. Tears also include oil, which helps to stop the water evaporating, and mucus, which helps the water to spread evenly across your eyeball.

✓ **FACT** **OR** **FICTION**

68 ake your fortune! Bring out the oil drills! Your body is covered in tiny oil wells. There are even oil glands along your eyelids. These tiny oil glands help to keep the eye moist. They are on the upper and lower eyelids and there can be between 50 and 70 in each eye. So many oil glands, but so tiny. Perhaps put away those oil drills. Not enough oil here to make any money.

✓ **FACT** **OR** **FICTION**

Freaky Fact or Fiction

69

Long words, short words. The weight or importance of a word is not measured by its length. It's the same with the body. A hip is a hip whether you call it so or use its proper name, *articulatio coxae*. Here are eight more body words that have just three letters. Big and small, we need them all. Eye, leg, arm, toe, jaw, rib, lip and gum.

✓ **FACT**   **OR**   **FICTION**

70 'Blink and you'll miss it' is a common expression. It usually means that something is very fast. Like a train, or a bird. Luckily most things move slower than that so you can see them. Just as well. The average person blinks about 12 times every minute, usually once every two to 10 seconds. That's 720 times an hour, 17,280 times per day and around 6,307,200 times per year. Imagine what you could miss!

✓ FACT OR FICTION

Freaky Fact or Fiction

71 Have you ever had a craving for eating clay, paper or other weird non-food things? It's different to when babies just put everything in their mouth. It's a medical condition where people want to eat things that are not food. It's called 'pica' (peek-ah). Doctors are not sure what causes it, but sometimes it happens in people with anaemia (with a lack of red blood cells, or red blood cells not working properly). Sufferers sometimes crave ice and say it tastes better than it does when they're well.

 ✓ FACT OR FICTION

72

Have a close look at your face in the mirror. Look at the faces of your family and friends. Any freckles there? Look at a baby or small child. Check your parents. Freckles don't usually appear in children under two years old. They tend to disappear in adults too. But even someone who has many, many freckles on their face won't have freckles on the parts of their skin that have not been out in the sun. (Like in their armpit, or on their bottom!)

 ✓ **FACT**   **OR**   **FICTION**

Freaky Fact or Fiction

73 Have you ever wondered why your nose gets runny when you cry? It's bad enough that your eyes overflow. Why does your nose have to do the same? Perhaps your nose just wants to be part of the action. When you cry, your nose begins to produce more mucus (snot or boogers). This mucus collects and then runs out of your nose, just like the tears run down your cheeks.

 ✓ FACT OR FICTION

74

ub-dub. Lub-dub. That's the sound of your heart beating. If you ever get a chance to listen through a stethoscope, that's the sound you'll hear. It's the sound of the heart working to pump blood around the body. From when you are born until you are an adult, your heartbeat gradually slows down until, on average, it happens about 70 times a minute. That's 4200 times an hour, 100,800 times a day and nearly 37 million times in a year.

✓ FACT OR FICTION

Freaky Fact or Fiction

75 An adult human heart beats around 70 times per minute. In general, the smaller an animal is, the slower its heart rate will be. The larger an animal the faster its heart rate will be. A canary has a heart rate of about 25 beats per minute. An elephant's heart rate is about 1000 beats per minute.

✓ FACT   OR   FICTION

76

Do you sometimes feel invisible? People with Cotard's syndrome do. Sometimes they think they have no heart, or no stomach. Sometimes it's so bad that even though they know they are walking around, they are sure they're dead. They continue to believe this, even when they are reassured that they are alive. Sometimes they are also convinced that the world doesn't exist anymore. Not a happy place.

✓ **FACT** **OR** **FICTION**

77 Health experts suggest that eating fish is a good thing to do. It's lower in fat than beef and pork and has other healthy benefits. And it's delicious. But even if you love the taste of fish, you probably would rather not smell like one. Particularly one that's died at sea and washed up onto the beach and is super-rotten and pongy. People with 'fish odour syndrome' smell like those fish. They have fishy breath and their sweat also smells like rotting fish.

 ✓ **FACT**   **OR**   **FICTION**

78

The first successful kidney transplant was performed by a medical team in Boston, USA, in 1954. The first successful heart transplant was performed by South African surgeon Christiaan Barnard in 1967. Since then medical science has discovered how to successfully transplant many different organs, including liver, bone marrow, cornea, lungs and more. Maybe one day they will be able to transplant every part of your body. Imagine!

✓ **FACT** **OR** **FICTION**

Freaky Fact or Fiction

Sometimes when people are very cold, their lips and fingers can appear blue. But the blue soon disappears when the person warms up again. Their skin colour returns to a normal pinky colour. The blue colour is not permanent. Unless of course you have argyria (ar-jir-ee-a). People with this condition have blue skin. Their gums, fingernails and toenails can also be blue. And it's nothing to do with the cold. It's caused by a build-up of silver inside their body.

 ✓ FACT OR FICTION

80

Where on your body would you find a philtrum? What do you do with a philtrum? How many of them are there? Well, there's only one and it's on your face. It's the vertical groove above your upper lip, just below the middle of your nose. It's where some men grow a tiny moustache. And it doesn't really do very much at all. It just is.

✓ **FACT** **OR** **FICTION**

81

Would you like to be taller? Almost instantly? Then head out of your house, out of your town, out of this world. The place you want is space. In space, you weigh nothing, and you become taller. All you have to do is finish school and university, join a space program and take off. Easy! In space, or anywhere else where there is no gravity, you could soon be about 3 per cent taller.

✓ **FACT**   **OR**   **FICTION**

82 Is it a bird? Is it a crocodile? Is it a dolphin? Is it human? Surely you can tell the difference! They're nothing alike. Perhaps not now, but if you look at the bones of these animals' forearms you may see some similarities. Many scientists believe they all evolved from a common ancestor. They may be different sizes and the proportions different, but the same bones are there. It's lucky we don't have to assemble our bodies from a kit. Imagine if you picked the wrong bits and ended up with short stumpy crocodile forelegs instead of forearms.

✓ **FACT** **OR** **FICTION**

Freaky Fact or Fiction

83

Imagine having a sibling that is with you for your entire life. Everywhere you go, they go too. Whenever they speak, you can hear them. That is what conjoined twins experience. Conjoined twins are identical twins who are physically joined when they are born. The most well-known conjoined twins are Eng and Chang Bunker, who were born in Siam (now Thailand) in 1811. They were joined at the lower chest, but other conjoined twins have been born joined at the hip, the stomach and even the head! Thanks to developments in surgery, it is now possible for some (but not all) conjoined twins to be separated.

✓ FACT OR FICTION

84

ust is just a bit of dirt, isn't it? Well, there may be soil in dust but there are also many other ingredients. House dust includes wool, cotton, paper fibres, fingernail clippings, food crumbs, pollen, foam particles, salt and sugar crystals, animal dander (dog and cat skin and fur), fungal spores and wood shavings! It also includes quite a lot of human skin flakes. Every day, breathing in all that stuff – it's almost enough to make you WANT to clean your bedroom. Almost.

✓ **FACT** **OR** **FICTION**

85

Dust mites are related to spiders and have eight legs. They are so tiny you can hardly see them, but they are there in their thousands. Their favourite food is flaked-off skin. Yum. Yum. They crawl around your bed and pillow. Delightful. Some people are allergic to dust mites, but for most people, dust mites don't cause any problems as long as you change your sheets regularly and vacuum every once in a while.

 ✓ **FACT** **OR** **FICTION**

86 Have you ever had a blister? Perhaps on your hand from swinging on the monkey bars, or on your foot where a shoe has rubbed. It's usually no big deal and though it might hurt for a little while, it soon heals and is forgotten. But there is a rare condition called epidermolysis bullosa (epp-ee-derm-o-ly-sis bull-o-sa) where blisters can never be forgotten. People with epidermolysis bullosa have very fragile skin and their blisters can be enormous and leave scars.

✓ **FACT** **OR** **FICTION**

87

Do you bite your nails? It's not the most attractive of habits. Think of all the germs that you swallow with the bits of nails. Even if you spit out the nails, some of the germs stay in your mouth. Like all habits, it's a hard one to break. It even has its own medical name. It's called onychophagia (on-ick-o-fay-jah). The word literally means finger-mouth.

✓ FACT   OR   FICTION

88 Melanin gives colour to your skin. Being in the sun causes your body to produce more melanin, to help protect you from the sun. But some people, those with albinism, have less or no melanin in their skin, hair and eyes. Animals with albinism will lack melanin in their feathers or scales. They will look white. People with albinism often have very pale skin and hair. But they do have good night-sight.

✓ **FACT** **OR** **FICTION**

Freaky Fact or Fiction

89

Some things are good and some things are bad. Ice-cream is good? Yes. Bacteria is bad? Well, yes and no. There are good and bad bacteria and your intestines are full of the good kind (if you are healthy). Your body needs the good bacteria to help break down food into bits that the body can use. In fact, an average adult carries about 1.5 kg (3.3 lb) of good bacteria in their intestines. Bad bacteria are the ones that make you sick. Just as well your body can tell the difference.

 ✓ FACT OR FICTION

90

ave you ever heard people talking about someone wearing 'rose-coloured glasses'? Usually it means that they are being too optimistic, or not realistic about something or someone. If you have cyanopsia (sy-an-op-see-a), however, you see everything with a blueish tinge. And if you have acyanopsia, you can't recognise blue at all.

 FACT **OR** **FICTION**

Freaky Fact or Fiction

91 Achromatopsia (ay-crow-ma-top-see-a) is total colour-blindness and is very rare. More common is colour-blindness to just one colour. If you had red colour-blindness you'd find it hard to tell the difference between red and green. If you were blue-blind, you'd struggle to tell the difference between blue and yellow. And if you were green-blind, you just wouldn't be able to see green.

✓ FACT OR FICTION

92   About one in every 10 boys and girls has colour-blindness. That means there could be more than one colour-blind child in every classroom. Children with red-green colour-blindness have trouble telling the difference between red and green. The condition can be hereditary, programmed in your genes before you are born. If both your parents carry a gene for colour-blindness there's a chance you will have it too.

 FACT   **OR**   **FICTION**

93

The biggest artery in the body is the aorta (ay-ort-ah). It is about 3 cm (about 1 in) in diameter. Make a circle with your thumb and first finger. It's about that big. The smallest blood vessels are the capillaries (cap-ill-ah-rees), which are only about one cell wide. If you put all the blood vessels together, end to end, they would reach twice around the world.

✓ FACT   OR   FICTION

94

The heart is very hardworking. Each minute it pumps 5 L (about 10.6 pt) of blood around the body. That's the total volume of blood for an average adult. So, that's 5 L a minute, 300 L an hour, 7200 L a day. That's more than 2.5 million L (660,000 gal) of blood pumped through the heart in one single year. That's about the volume of water in an Olympic-sized pool. That's a lot of pumping. Especially for a pump that is only the size of your clenched fist.

 FACT **OR** **FICTION**

Freaky Fact or Fiction

95 Humans and sharks – they're very different creatures. Instead of legs, sharks have fins. Sharks have tails; humans don't. Sharks live in water; humans live on land. Sharks have gills; humans have lungs. But there is one way, at least, where sharks and humans are similar. The cornea of the shark eye is very similar to the human cornea. In fact, it is so similar that shark corneas have been transplanted into humans. Corneas help humans (and sharks) to see.

 ✓ FACT OR FICTION

96

ACHHHOOOOO! You closed your eyes! ACHOO! ACHOO! You closed them again. Next time you sneeze, see if you can keep your eyes open. It's almost impossible. That's because a sneeze is a reflex that affects more than just your nose. Your face muscles, throat and chest are also involved. First something irritates your nose. Then you take in a big breath ... a ... ahh. Then you breathe out very quickly and violently to clear your nose. ACHOO!

 ✓ **FACT**   **OR**   **FICTION**

97 **W**hen you sneeze, air and other bits (hopefully including whatever caused the sneeze) are exploded from your nose. Fast. Faster than you can run. Faster than a speeding bicycle. Faster than a cheetah sprinting. The speed of a sneeze can be up to 150 km/h (about 93 mi/h). If your sneeze was a cyclone, it would be classed as a Category 1: gale. Damage to crops and trees. Might drag boats from their moorings.

 ✓ **FACT** **OR** **FICTION**

98

Have you ever sleepwalked? Chances are you don't remember it. Sleepwalkers seldom do. But they can sit up and climb out of bed. Sleepwalkers sometimes get dressed. They may walk with their eyes open and can find their way around corners. You can even have a conversation with a sleepwalker, although they may not make all that much sense. Amazing. And they have no memory of any of this the next day.

 ✓ **FACT** **OR** **FICTION**

99

Frog. Snake. Human. That's how your heart develops. Huh? When you were growing in the womb, your heart started out as just a collection of cells, then a tube. For a while it looked like a frog's heart, with only two chambers. Next it developed a third chamber and began to look more like a snake's heart. Final stop was four chambers and then your heart looked just like it should. Four chambers, all contributing to pumping your blood to your head, to your fingers and your toes.

✓ FACT   OR   FICTION

100 It seems that everything in the body has a name. Even the mashed-up food that leaves your stomach on the way to the small intestine has a name. It's called chyme (kyme). Chyme is a thick liquid that sits in your stomach for hours. Then it is spurted into your intestine in little bursts. That soupy mush used to be your favourite food. It doesn't look so yum now!

✓ **FACT** **OR** **FICTION**

Freaky Fact or Fiction

101 Sphincters (sfink-ters) are like rubber bands. They are ring-like muscles that tighten and close around different tubes in your body. They tighten to close and relax to open. There are sphincters at the top and bottom of your stomach – a bit like having ties at each end of a balloon. There are also sphincters around the bottom of the large bowel and at the end of the urethra. Lucky really, because if there were no sphincters at these two places, you'd have to wear a nappy for your entire life!

✓ **FACT** **OR** **FICTION**

102

Sphincters (sfink-ters) are ring-like muscles that open and close to control what goes in and what comes out of certain sections of your body. One sphincter you can easily see is your iris. If someone shines a bright light into your eye, the sphincter shrinks or tightens. If you are trying to read in bed after lights out, your iris will relax and let in as much light as possible.

✓ **FACT** **OR** **FICTION**

Freaky Fact or Fiction

103

Yawn. There's nothing nicer than crawling into bed at the end of a long, busy day and sleeping. First you close your eyes, then relax your body and gradually, gently, your body drifts off to dreamland. It doesn't work like that for people who experience narcolepsy (nark-o-lep-see). They can fall asleep very suddenly. For example, you might be in the middle of telling them the most exciting story and they will fall asleep. On their feet. Then, a few minutes later, they'll wake up just as suddenly, and can tell you exactly what you were saying while they were asleep.

✓ **FACT** **OR** **FICTION**

104 It's hard to get moving some mornings. You just want to turn over and slip back into that dream about being on a beach in the sun. But you can't. You have to get up to go to school and eventually that's just what you do. But if you have narcolepsy (nark-o-lep-see) you can't always do that. Get up, that is. Some people with narcolepsy wake up in the morning but can't move anything. They are paralysed, although their brain is awake. Scary. Luckily it doesn't usually last for long and they too can get up and get ready for school!

✓ **FACT** **OR** **FICTION**

Freaky Fact or Fiction

105

Laughing is great fun. From the tiniest chuckle to the biggest, shake-the-walls guffaw, laughing helps you to feel better. But have you ever laughed so much you couldn't stand up? Have you ever fallen over laughing? People who do may have dogaplexy, a condition where you briefly lose control of your muscles and fall over.

✓ **FACT**   **OR**   **FICTION**

106 **H**as anyone ever told you to go to bed because you are tired and grumpy? When you are extremely tired, it's hard to make sense of anything. It can seem too hard to brush your teeth or even to put on your pyjamas. Extreme tiredness from lack of sleep is called sleep deprivation (deh-pri-vay-shun). Some other symptoms of sleep deprivation include blurry eyes, hard-to-understand speech and being confused. Nighty-night. Off to bed with you!

✓ **FACT** **OR** **FICTION**

Freaky Fact or Fiction

107

Ouch! Sometimes when you fall over or fall out of a tree, or fall out of a tree and then fall over, you scrape skin. Sometimes it feels like you've taken off many layers of skin. Some weeks it feels like you've lost almost all your skin. The epidermis (epp-ee-derm-us) is the outer layer of your skin and even without falling over, it comes off all the time. It may take a month or more for an epidermal cell to grow, die, flatten and then flake off all by itself. Of course you lose skin instantly if you fall off your bike.

✓ **FACT** **OR** **FICTION**

108

How much can you eat? Do you ever feel so full you think you might burst? As if you'd swallowed a watermelon? There is a medical condition called watermelon stomach. You don't have to eat a watermelon to have watermelon stomach. In fact, doctors don't really know what causes it. They just know that the inside of your stomach looks stripy, like a watermelon!

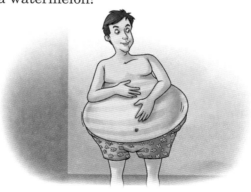

✓ **FACT** OR **FICTION**

Freaky Fact or Fiction

109

A phobia is a fear that affects how you live. Some uncommon phobias are: harpaxophobia (fear of robbers); sitophobia (fear of food); frigophobia (fear of being cold); pogonophobia (fear of beards); chaetophobia (fear of hair); stasophobia (fear of standing); phronemophobia (fear of thinking); and acarophobia (fear of itching caused by insects or other bugs).

✓ **FACT**   **OR**   **FICTION**

110

Trichotillomania (trik-o-til-o-may-nee-a) is a big word. Can you guess what it means by breaking down the word? It sounds almost funny or tricky. But trichotillomania isn't tricky or funny. People with this condition feel they have to pull out their hair. They pull out hair from their head, their eyelashes, their eyebrows – from anywhere on the body. To cover up the bald patches, they sometimes wear wigs or false eyelashes.

✓ **FACT** **OR** **FICTION**

111

What's the strangest thing you've ever eaten? Ox tongue? Lamb's brains? Liver? Crickets? Snake? What about hair? Cats and dogs swallow hair, though it's more by accident when they are grooming themselves. But people? People don't eat hair do they? Well yes, sometimes they do. And like cats and dogs who have fur balls in their stomach, people can build knotted balls of hair in their tummies. A human hair ball is called a trichobezoar (try-ko-bez-o-ar). The human body can't digest it and will soon push it on out.

✓ **FACT**   **OR**   **FICTION**

112

Surgeons use scalpels (very sharp-bladed knives) to make their incisions (cuts). After they've finished whatever they were doing, they need to close the wound so it will heal well and also to keep out any bacteria that might cause infection. Sometimes they use sutures (stitches) to keep the wound closed and sometimes they use a dressing called a caterpillar. It's called this because it's long and narrow, like a caterpillar.

✓ **FACT** **OR** **FICTION**

113

When is an uncle not a good uncle? When it's a furuncle. A furuncle is also known as a boil. It's a hard, infected lump just under the skin. It hurts! Worse still, it's full of pus and won't heal until the pus is out, out, out. When is an uncle an even worse uncle? When it's a carbuncle. A carbuncle is a collection of furuncles all connected under the skin. All those little volcanoes! Stand back! They're about to explode!

✓ **FACT**   OR   **FICTION**

114

Your brain is plastic. Actually, it's neuroplastic, which isn't quite the same thing. Break the word into its two parts and look at the meanings. 'Neuro' is to do with nerves. 'Plasticity' means being able to be moulded or changed. So neuroplasticity means nerves that can be changed. If you have an accident, nerves that once did one job can change and take the place of the nerves that have been damaged. Very clever.

✓ **FACT** **OR** **FICTION**

Freaky Fact or Fiction

115

When is a ghost not a ghost? When it's a phantom. Or more accurately, when it's a phantom limb. People who have had a leg amputated (cut off) sometimes 'feel' their leg is still there. They know it's gone, but the nerves that used to connect to the amputated leg sometimes twinge, sending messages to the brain that their ankle is sore or itchy. It's called phantom limb syndrome.

✓ FACT OR FICTION

116

Gangrene isn't green. It's black.
Dry gangrene happens when blood can no longer reach a particular part of the body, like toes or fingers. When there's no more blood, the toe will slowly dry up and turn black. Then the part of the toe that's affected may just fall off and the rest of the toe will be fine. There is another type of gangrene. It's called moist gangrene. It tends to spread quite quickly once it has begun and can be very smelly.

✓ **FACT** **OR** **FICTION**

Freaky Fact or Fiction

117

Sit still! Stop wriggling! Agggh! How can you sit still when your legs won't let you? They want to keep moving around. You might be surprised to know that doctors understand. They even have a name for it. It's called 'restless legs syndrome'. At its worst, it can stop you from sleeping at night and make you feel sleepy during the day.

✓ **FACT**   **OR**   **FICTION**

118

Have a look at your legs. Look at the part from your hip to your knee. In most people, it's straight, or fairly close. Now, look at the bit between your knee and ankle. It's also more or less straight. If you had the disease called rickets, your bones might not be so straight. In rickets, the bones soften and can become quite curved. This bone-softening disease, caused by a lack of Vitamin D, only happens in children.

✓ FACT OR FICTION

Freaky Fact or Fiction

119

Tonsils are at the back of your throat. They are almost invisible when they're doing what they're supposed to. Their job is to make sure germs stay out of your stomach and lungs. But every now and then, tonsils seem to decide they want to be noticed. When they become infected and swell up, they can be very uncomfortable. Then they might have to be removed.

 ✓ **FACT** **OR** **FICTION**

120

Bite. Bite. Bite. Itch. Itch. Itch. Fleas are a real pest if they get into your house and start biting. But soon enough the itching stops and the bite is gone. Unless, of course, the flea is the jigger flea (also known as a chigoe flea or sand flea). They're nasty little creatures that burrow in between your toes and lay their eggs in your skin. They also feed on your blood. Nasty little vampires. They can cause all sorts of horrible infections. Most people are lucky and will never see these little blighters. They are only found in tropical climates, such as South and Central America and the West Indies.

✓ **FACT** **OR** **FICTION**

Freaky Fact or Fiction

121

Wouldn't it be wonderful if you could trade in the bits of you that weren't working properly and get new ones? Imagine if you could trade in those tired eyes after a busy week. There are some bits of the body that can be traded in, although it's not quite as easy as changing your socks or shoes. Some parts of the body can be replaced with artificial bits, and others are replaced by transplants. It is possible to get artificial shoulders, knees, hips, toe joints and valves. Transplants include body parts such as heart, lung, bone marrow and tonsils.

 ✓ FACT   OR   FICTION

122

Fall off a swing. Slip through the branches of a tree. Fall over. Chances are you'll get a bruise and maybe some scrapes, but you'll usually manage to mend without breaking a bone. Usually. Bones give shape to your body. Without bones, you'd collapse in a lumpy puddle on the ground. Bones are strong, hard and solid.

 ✓ FACT OR FICTION

Freaky Fact or Fiction

123

An organ is a musical instrument a bit like a keyboard or a piano. Bits of your body are also called organs, even though they don't make much noise (well, some, like your stomach, can!). The heart is an organ. So is the liver. But the biggest, heaviest organ of all is the skin. An average adult's skin weighs about 5 kg (11 lb) and, if stretched out, would cover most dining tables at 2 m² (21.5 ft²).

✓ **FACT**   **OR**   **FICTION**

124 All arteries pump blood that has been through the lungs and is carrying oxygen. All veins carry blood that's used up most of the oxygen. That's one way you can tell the difference between the blood vessels, by the colour of the blood in them. The arterial blood will be a brighter red, the venous blood darker.

 ✓ FACT OR **FICTION**

125

What is a heart attack? Does something attack the heart? A heart attack is also called a cardiac arrest. How does that make sense? Nothing, or no-one puts handcuffs on the heart. In this sense, 'arrests' means 'stops', and that's what happens in a cardiac arrest. The heart stops because the blood supply to a coronary artery has stopped. This means there is no blood getting to part of the cardiac (heart) muscle. The heart muscle can't function without blood, and stops.

 ✓ FACT OR FICTION

126

Sometimes when you hurt or scratch yourself there's hardly enough blood for a bandaid. Sometimes there's only one tiny drop of blood. It might be small, but just think about what's in that one single little drop: about five million red blood cells, plus thousands of white blood cells and platelets, all carried in a plasma soup. How tiny must the bits of your blood be?

 ✓ **FACT**   **OR**   **FICTION**

127

There are arteries and there are veins. They run side by side like twin rivers, even though arteries go one way and veins go the other. But there's another system that travels alongside the arteries and veins. It's called the lymphatic (lim-fat-ic) system. In places, such as under your arm and in your throat, there are small clusters of cells called lymph nodes. They look like tiny beans. Their job is to destroy germs and keep you well.

 ✓ **FACT**   **OR**   **FICTION**

128 It is the job of some nerves to help you feel things. Think about where you feel things most. Fingers. Particularly the tips of your fingers. Each of your fingers has four nerves branching out like trees folding across the tip so they are super-sensitive to all sorts of touch. But not everywhere in your body has quite so many nerves overlapping. That's why you feel things with your fingers and not your elbow. Or your nose tip. Or your bottom.

✓ **FACT** **OR** **FICTION**

Freaky Fact or Fiction

129

Earwax is that yellowy gooey stuff in your ears. Its only purpose seems to be to make your mother/father/grandmother/grandfather/other family adult say, 'Go and clean your ears!' But, like most of the other things in your body, it has a purpose. Earwax, that is. Earwax helps to keep your ears clean by collecting all the dead skin cells, dirt and fallen-out hairs, and slowly carrying them to the outside of the ear canal. That's why everyone says not to stick things into your ear to get the wax … because the wax will eventually come to you.

 FACT   OR   **FICTION**

130

Memory is a bit like sorting out your clothes for the year. Short-term memory is for the clothes you're going to wear today, this week, this month. They're easy to get to. Long-term memory is like putting your coats and hats and gloves away in another cupboard in spring, because you won't need them until next winter. Another part of memory, just as important, is retrieval, or recall. Thinking about WHERE you put things, so you can find them if there's a sudden cold snap, is important. That's recall.

✓ **FACT** **OR** **FICTION**

131

Bones are very strong but they do sometimes break. And if they do, generally the body is very efficient in healing the break, or fracture, even if it needs a plaster cast support for a while. First new bone cells are laid down around the fracture, almost like scaffolding. Then bone cells arrange themselves across the fracture to rebuild the bone and its strength. Then, when the bones are finished healing, the body pulls down the scaffolding and carries it away. But because the rebuilt bone is slightly denser, it will always be able to be seen on X-ray, even years later.

 ✓ FACT OR FICTION

132

Have you ever had a fever? You shiver and sweat and feel sore all over. It's horrible and all you want to do is curl up in a ball and try to sleep. But it's not even easy to sleep. Your sheets crumple and twist as you toss and turn to find a comfortable position. You are not interested in food, and hardly interested in drinking. This is what infections do to you. There is nothing good about a fever.

✓ **FACT** **OR** **FICTION**

Freaky Fact or Fiction

133

Long-distance swimmers sometimes smear their bodies with thick layers of vaseline or other greasy stuff to help protect their skin. With the same purpose, a baby growing in the womb becomes covered in a layer of greasy whitish stuff called vernix. This helps to keep their skin soft but not soggy. After all, they spend a long time floating in liquid, much longer than even the longest long-distance swimmer will.

✓ **FACT**   **OR**   **FICTION**

134

Some animals, like cats and dogs, have litters of babies. But not humans. They usually only have one baby at a time. Usually. Sometimes two. Occasionally three. Rarely four or more. Twins can happen in two ways. Sometimes the one fertilised egg splits into two. When that happens the twins born will be identical. The other way twins happen is when two eggs are released at the same time and fertilised by different sperm. These twins will only look as alike as other children in the family.

 ✓ FACT   OR   FICTION

Freaky Fact or Fiction

135

The smallest bone in the body is the stapes (or stirrup) bone in your middle ear. It's about 3 mm (0.1 in) long and weighs about 3 mg (0.0001 oz). It's hard to even imagine a bone that small. A teaspoon of sugar weighs about 5 g. So one teaspoon of sugar weighs as much as 1.667 of these stapes bones. Can you work out how many grains of sugar that would be?

✓ FACT   OR   FICTION

136

Has anyone ever called you an airhead? If they did, they probably meant that you weren't being very clever. But you do have pockets of air in your head. And they are very cleverly designed. Your middle ear is like a tiny air-filled room. It even has windows. Three tiny bones cross this 'room'. They are the anvil, the hammer and the stapes (stirrup).

 ✓ **FACT** **OR** **FICTION**

137

The largest bone in the body is the femur. That's the bone that connects your pelvis with your knee. It can grow to about 50 cm (about 20 in) long. It's also very strong. At the top it forms a ball-and-socket joint with the pelvis. Ball up one fist and cup your other hand around the fist. Now move your fist around. You can move the fist around without leaving the cupped hand. That's how your hip moves. At the other end of the femur the joint works differently, mostly just backwards and forwards.

 ✓ FACT   OR FICTION

Ah, roast chicken! Delicious! Do you fight your brothers and sisters for the most delicious part? It's probably not the bones or the cartilage. What's cartilage? Have a look at the drumstick. At each end there's a whitish sort of cap over the end of the bone. That's cartilage. Of course it's changed by being cooked, but it's a bit similar to the cartilage that covers the ends of the bones in your body. Cartilage may look very thin, but it's very good at protecting your bones.

✓ **FACT** **OR** **FICTION**

139

In gymnastics, one of the disciplines is walking along a balance beam. Actually, once you've walked along it, you'll learn to do many other things, but they all depend on balance. Your balance centre is in your inner ear. It helps you to know when you are upright, when you are lying down and when you are anywhere in between. It also helps you to quickly move your muscles so you don't fall off the balance beam.

 ✓ **FACT**   **OR**   **FICTION**

140

hee-ewww! Can you smell that? If you can, it's because of the olfactory sensors inside your nose. They are long thin cells that end in delicate hairs called cilia (sill-ee-ah). They pick up smells from chemicals in the air that you breathe in. The information is sent to your brain and you recognise whether it's a good smell or a bad smell. Humans can sense more than 300 different smells, but they are grouped into six main types: fruity, flowery, resinous, spicy, foul and burned.

 FACT **OR** **FICTION**

Freaky Fact or Fiction

141

I f someone says 'chocolate' to you, can you sometimes almost taste its deliciousness? In someone with synesthesia (sin-es-thee-sha), things work a little differently. They may connect words with visual sensations like colour. For example, someone with a colour synesthesia might see each letter of the alphabet as having a different colour. Someone with a taste synesthesia might think that a particular word, for example 'tomorrow', might taste like spinach.

✓ **FACT**　　　　**OR**　　　　**FICTION**

142

If you're going somewhere you've never been before, it's handy to have some directions or a map. Imagine if

you had maps on your arms that you could just call up when you need them. What about on your tongue? Then you'd need to have a mirror to read them, but at least you could keep the map out of sight. Odd? Yes. But there is a condition called 'Geographic Tongue' which causes patches on your tongue. It looks like a map. Who knows where it would lead you!

✓ **FACT** **OR** **FICTION**

Freaky Fact or Fiction

143

Cheese belongs on toast or in a sandwich or even on top of pasta. Anywhere but on your feet. And people with bromidrosis (broh-me-dro-sis) probably wish that's the only place cheese belonged. But sometimes the sweating that comes with this condition is so bad that bacteria start to grow in the soggy skin. These bacteria release a gas that smells similar to a ripening cheese. That's why this condition is sometimes called 'cheesy feet'.

✓ **FACT**   **OR**   **FICTION**

144

Can you roll your tongue? How many of your friends can roll theirs? More than half of your friends should be able to do it, according to studies that have been done. It's a genetic thing. You either can do it or you can't. It's one of the things you can inherit from your parents like the colour of your eyes, or the colour of your hair.

✓ **FACT** **OR** **FICTION**

Freaky Fact or Fiction

145

Did you know you have scavengers in your blood stream? In the wild, scavengers are those animals that don't hunt, but eat what they find or what others have left behind. In your bloodstream, scavengers called macrophages (mak-ro-farj-z) eat germs and other things that shouldn't be in your blood. They are like a gardener pulling out weeds. Gobble, chomp, gobble. Gone.

✓ **FACT** **OR** **FICTION**

146

Macrophages also get rid of blood cells that are worn out or broken, a bit like a gardener pulling out old or sick plants. Sort of. Imagine a macrophage (mak-ro-farj) as being a bit like a jellyfish with no tentacles. It wobbles up to an old worn-out, not-working-any-more, out-of-shape red blood cell. Slowly it surrounds the red blood cell until all you can see is macrophage. Keep watching. You might not see much happening, but inside, the macrophage is slowly digesting the red blood cell. Delicious!

 FACT OR FICTION

147

High or low, sinuses can cause trouble. Air sinuses are air-filled spaces in the bones next to your nose. If you've been on a plane, you might already know this, because sometimes these sinuses can hurt when you are taking off or landing. Especially if you can't clear your ears by swallowing. Sinuses can also cause trouble if you are scuba diving. This pain is called 'sinus squeal', because you can hear a high-pitched squeal as the body tries to decrease the pressure in the sinuses.

✓ **FACT**   **OR**   **FICTION**

148

She says it's hot, he says it's cold. Surely they can't both be right? There are spots on the skin that respond to heat, or to cold. But usually it's one or the other, not both. However, it can happen that if something very warm is put on a spot that usually detects cold, it will tell the brain that it's cold. When it's not. Confusing? That's why it's called paradoxical (paa-ra-dox-ee-cul) cold, which means not-quite-as-it-should-be.

 ✓ **FACT** **OR** **FICTION**

Freaky Fact or Fiction

149

You have a seahorse in your brain. But that's not what it's called. It's called the hippocampus and it's shaped like a seahorse. 'Hippo' comes from a word that means horse, and 'campus' comes from a word that means sea monster. The hippocampus helps to store long-term memories. Long-term memories are those memories from when you were little or from last week, or last month. Perhaps somewhere in your hippocampus there is a memory of seeing a seahorse in an aquarium or in the ocean.

 ✓ FACT OR FICTION

150

What's the opposite colour to blue? Or to green? Don't know? There's an easy way to tell. You'll need really good light, so don't try this in your bedroom after lights out. Stare at a patch of blue for 20 to 30 seconds. Then stare at something white. You'll see a patch of yellow. Now stare at a patch of green for 20 to 30 seconds and again look at something white. You'll see a patch of a dark pinky colour. Weird. The colour you see on the white is called an 'afterimage' and it's the opposite of the first colour you saw.

✓ **FACT** **OR** **FICTION**

Freaky Fact or Fiction

151

Have you ever heard anyone talking about having a fussy palate (pal-ett)? They usually mean they are fussy eaters, and they don't like lots of foods. But what exactly is a palate? The word looks like a cross between a plate and a palace, but it is neither. The palate is the roof of your mouth and it has two sections. One part is hard and doesn't move at all. The other part is soft and moves up and down when you swallow or when you suck something, like when you suck on the straw in a strawberry thickshake. It stops food (or thickshake) going up your nose. A useful thing!

 ✓ **FACT**   **OR**   **FICTION**

152 Cartoon characters open their mouth very wide when they cry or when they scream. Sometimes you'll see a little dangly bit at the back of their throat. It might wobble wildly, or just hang down like a big water drop. Sometimes cartoons will also show it vibrating when someone is snoring or singing. That dangly thing is called a uvula (you-vue-lah) and we all have one. It's like your appendix – it does nothing at all.

 ✓ FACT OR FICTION

Freaky Fact or Fiction

153

The smallest finger on each hand is sometimes called your pinky. Sometimes this is spelled pinkie. The name has been around for years and has nothing to do with the finger's colour. It's called the pinky because it's the smallest. It comes from an old Dutch word for small, '*pinck*'.

✓ FACT   OR   FICTION

154

The body is full of so many useful bits and pieces, there seems to be no room for bits that don't do anything. But then there's the appendix. The appendix is a little finger-sized tube that sits near where the small intestine meets the large intestine. Nobody seems quite sure what it's for. Some think it is left over from when humans had a different diet. Others think it might help grow more bacteria for the bowel. But for a tiny little, do-nothing organ, it can cause plenty of trouble if it gets blocked.

 ✓ **FACT**   **OR**   **FICTION**

155 How many times have you been told that sugar is bad for you? Too much processed sugar can cause all sorts of trouble. In your mouth, processed sugar can damage your teeth; in your body it can give you a short-lasting energy boost. But sugar also occurs in milk and plants. These forms of sugar are essential to keep your body going. The sugars are broken down in your body until they are in a form that can be used by your tiniest cells.

 FACT **OR** **FICTION**

156

Insulin is a chemical made in your body to transport sugars into your cells. Insulin molecules are a bit like a bridge from your bloodstream to your cells. Insulin is made in special parts of the pancreas called the Islets of Langerhans (eye-lets of lang-err-hanz). People with diabetes have problems with their insulin. It might sound wonderful, but too much sugar in your blood can mean not enough in your cells and that can cause all sorts of not-so-fun problems. People with diabetes can also have trouble with a lack of sugar in their blood.

✓ **FACT** **OR** **FICTION**

Freaky Fact or Fiction

157

The iris (eye-ris) is the coloured part of your eyeball. It's made of muscles that contract and relax to control the amount of light that reaches the back of your eye. Iridology (ee-rid-ol-oh-jee) is the study of the iris. Iridologists (ee-rid-ol-oh-jists) are people who study iridology. Iridologists believe they can tell you about the health of your organs (like your kidneys or liver) by closely studying your iris.

 ✓ **FACT** **OR** **FICTION**

158

uack quack! It's a duck! Well, yes, ducks do quack. But 'quack' is also the name given to someone who pretends to be able to make sick people well. Quacks promise great things and often charge a lot of money, without any proof or evidence that they can really help. Quacks have been around for a long time, selling elixirs and other supposed medicines. The word quack here comes from an old Dutch word meaning 'boaster'. In this sense, a quack is boasting that their medicine can do more than it probably can.

 ✓ FACT **OR** **FICTION**

Freaky Fact or Fiction

159

Your eyeball is also called a globe. It is about the same size as a large marble. Some marbles look a bit like eyes, but your eye isn't much like a marble really, apart from the round shape. For a start your eye isn't made of glass. And it weighs much less than a marble. A fully grown human eye weighs about 30 g (1 oz). That's about the same as six teaspoons of sugar.

✓ **FACT**   **OR**   **FICTION**

160

The eye has a hole in the middle of it.
Oh, no! Oh, yes. Each eye has a hole right through the centre. This hole is called the pupil and it lets light in. The light hits the back of the eye and information is relayed to the brain and back again. This happens so fast that you're not even aware of it. But you can sort of see it when you see a photo where there is 'redeye'. 'Redeye' is the flash light bouncing onto your retina and being caught in the photo image. Makes you look weird too.

 ✓ **FACT**   **OR**   **FICTION**

161

Open your eyes. What can you see? Everything. You can see so much because there are rods and cones at the back of your eye. That's the name for the cells that decode what you see. Cone cells help you work out the shape, size and brightness of what you see. Rod cells tell you about the colour and detail of what you're looking at. There are about seven million cone cells and 130 million rod cells in each eye.

 ✓ **FACT**   **OR**   **FICTION**

162

Do you have a magnifying lens? Can you see how it's thicker in the middle than at the edges? If you have a bright sunny day you can burn a tiny hole in a leaf by focusing the light of the sun through the magnifying glass on to a single point on the leaf. The lens of your eye is also shaped to focus light. But instead of burning, it focuses light rays through the pupil and onto the retina so you can see.

 FACT OR FICTION

Freaky Fact or Fiction

163 Movies and cartoons sometimes show people with artificial eyes. A man might take his artificial eye out and it might roll away, setting up a comedy sketch where he keeps chasing it. But modern artificial eyes aren't actually round. They look a bit more like a fried egg, with the iris and pupil where the yolk of the egg would be. They are shaped to fit perfectly. Perhaps someone you know has an artificial eye. I wonder if you can tell.

✓ **FACT** OR **FICTION**

164

The medical world is full of long words. Many look impossible to pronounce, and even harder to understand. But many of them can be broken down into smaller words, which makes it easier to understand what they mean. Here's an example: 'hypereosinophilic'. 'Hyper' means 'more than' or 'too much'. So someone who is HYPERactive is very active. An 'eosinophil' is a white blood cell. And the 'ic' bit? It just turns the word into an adjective. So someone who is 'hypereosinophilic' has too many white blood cells. Easy as ABC.

✓ **FACT** **OR** **FICTION**

Freaky Fact or Fiction

165

The body is amazing. But it's not amazing enough for some people. They want more. Many people pierce their ears so they can wear earrings. Others pierce other parts of their face and body. Amazing! One woman, Elaine Davidson, who lives in Edinburgh, Scotland, decided she wanted to be even more amazing. She first set the record for body piercings in 2000 with 462 piercings, and by August 2001 she had 720 piercings. By January 2010 Elaine claims to have more than 6000 body piercings! That's really amazing!

✓ **FACT** **OR** **FICTION**

166

Tattoos have been around for a long time and have been used for many different reasons. But most people only have one, or a few. Not New Zealand – born Lucky Diamond Rich! He had colourful designs tattooed all over his body. But that wasn't enough. He then decided to have all his skin tattooed black. Even inside his ears and between his toes. Enough? Nup. More, he said. He now has white tattoos on his black tattooed skin, and then coloured tattoos on top of that!

 FACT OR **FICTION**

167

What makes you laugh most? Something you hear or something you see? Most likely it's a mixture of both. Humour, or what you find funny, is tricky to define. What makes you laugh may not make your sister laugh. Imagine... you see your sister almost fall, but do an odd dance instead before she gets her balance. You might laugh; she probably won't. You might laugh out loud at your favourite comedy program, but your parents might not find it at all funny. But what you can agree on is that laughing makes you feel good.

 FACT **OR** **FICTION**

168 Laughter is a reflex, just like vomiting and blinking. It involves the contraction of 15 particular facial muscles and a change to your breathing. The effect of laughing can be shown by using 'electrical stimulation' of a particular muscle (not something to do at home!). Depending on how much stimulation there is, the electricity can cause anything from a faint smile to a wide grin. And you thought it was something that just came naturally!

✓ **FACT** **OR** **FICTION**

169

Are you a faster runner than your father? Can you eat more than your brother? We all like to be good at something. Mostly it doesn't matter all that much who wins or is best. Most of us can do most things. But some people want to be known for being the best at what they do, particularly if it's something quite unusual. That's why we have the *Guinness World Records* book. Then if you are the best in the world at carving pumpkins, or eating cockroaches, everyone can read about your world record.

 ✓ **FACT**   **OR**   **FICTION**

170

Some days you feel so tired it can seem like you are 100 years old. Two hundred years old. A thousand years old. It's hard to lift your feet to walk, or to do anything, particularly clean the bedroom or wash dishes. Children with progeria (pro-jeer-ee-a) have better reasons than anyone to feel old. They get older much faster than the rest of us. By the time they are 10 years old, they can start to lose their hair, and their skin begins to wrinkle.

✓ **FACT** **OR** **FICTION**

171

Some things are easier to remember if you do them over and over. Your handwriting gets better with practice, as does your ball-throwing. Spelling is another skill that gets better if you do it over and over. But it's not always a good thing to do the same thing over and over. In OCD (obsessive-compulsive disorder) people do the same thing over and over again to help them feel better. It doesn't always work though. One common OCD behaviour is washing hands over and over again, but still not feeling your hands are clean.

✓ **FACT**   **OR**   **FICTION**

172 Each red blood cell looks a bit like a bagel with the middle filled in a little. It's thicker around the edges and thinner in the middle. There are between four million and six million in every microlitre (0.00003 oz) of blood. Each red blood cell lives for about 120 days (about four months) before it reaches retirement age. Then it changes shape. Instead of looking like a bagel, it looks more like a ball. It can't get into the small blood vessels then and will be removed by scavenger cells.

✓ **FACT** **OR** **FICTION**

Freaky Fact or Fiction

173

Police will often 'dust for fingerprints' when they investigate a crime. This is because fingerprints are unique. No two people have the same fingerprint. If they can find a fingerprint, they can sometimes find the criminal. Fingerprinting is also called dactylology (dak-till-oll-o-gee). People who study fingerprinting are called dactylologists.

✓ **FACT**   **OR**   **FICTION**

174

You've probably heard about hormones, those chemicals in your body that go nuts at puberty. But have you heard of pheromones (fer-o-moans)? Dogs and cats often use the pheromones in their urine to 'talk' to other dogs and cats. That's why dogs want to smell where other dogs have done a wee or a pee. Humans have pheromones too – in their skin. Scientists are still working out what they do, but think that we can sometimes recognise each other by smell.

✓ **FACT** **OR** **FICTION**

175

Oops! That could have been nasty. Have you ever nearly had an accident? Your body tingles as adrenalin (ah-dren-ah-lin) surges to your fingers and toes. It's quite a weird feeling. Imagine, though, if you nearly died. People who have near-death experiences sometimes talk about feeling like they were floating above their body watching themselves, or that they were travelling down a tunnel towards a bright light. Scientists aren't sure how or why this happens, but it's not uncommon.

✓ **FACT** **OR** **FICTION**

176

Mum, I feel sick… Have you ever pretended to be sick because you didn't want to go to school? Parents can usually tell when you're pretending. But not if you have Munchausen syndrome. People with Munchausen (munch-house-en) syndrome are very good at pretending that they have something wrong with them. They can often convince nurses and doctors and even end up in hospital. Sometimes they can be so convincing that they have operations they don't need. These people are unwell, but it's their mind that needs help, not their body.

 FACT **OR** **FICTION**

177

Keyholes are perfect for keys. Of course they are – that's what they're for. But 'keyhole' is also the description used for a particular kind of surgery. In keyhole surgery, the surgeon will make small holes instead of a big cut. They will use a little telescope and instruments on long handles to find where the problem is and then fix it. Patients heal much quicker from keyhole surgery, partly because the wounds are smaller.

✓ FACT   OR   FICTION

178

The instrument that a doctor will use to listen to your lungs and your heart is called a stethoscope (steh-thuh-scope). It has two earpieces at one end and at the other there is a two-sided listening device. One side, an open cone, will transmit the low-pitched sounds, and the other, which has a flat covering, is good at picking up higher-pitched sounds. The doctor will use both to check out just what your heart and lungs are up to.

✓ **FACT** **OR** **FICTION**

179 If you're sick, you might go to visit your local doctor, or general practitioner (GP). If you need special treatment, you might be referred to see a specialist doctor who has extra training. The specialist doctor you might see if you have problems in your ears is an otorhinolaryngologist (otto-rye-no-lar-ing-ol-oh-jist). Try saying that three times out loud! 'Oto' means ears, 'rhino' relates to your nose, and 'laryngo' is to do with your throat. The three are connected and part of one medical specialty. Most people call them ear, nose and throat specialists or ENTs. Much easier to say.

 ✓ FACT OR FICTION

180 common health check is to check a person's blood pressure. Blood pressure changes can indicate that something is wrong inside the body. The doctor will use a sphygmomanometer (sfig-mo-ma-nom-e-ter), sometimes in combination with a stethoscope. The doctor listens to the pulse near the elbow, then inflates a cuff around the upper arm until the pulse stops (measurement one). Then the air is slowly let out of the cuff and the doctor will listen to when the blood flow starts again (measurement 2). These two measurements can be high or low, or normal. Normal blood pressure for an adult is 120/80.

 ✓ FACT OR FICTION

Freaky Fact or Fiction

181

Glue ear is called glue ear because that's what the waxy stuff that builds up in your middle ear looks like. This glue can sit against your ear drum and sometimes it will affect your hearing. What? What did you say? Usually glue ear causes few problems. Sometimes, if it continues too long, doctors will put little drain tubes called grommets in to keep the fluid from building up and affecting your hearing.

✓ **FACT** OR **FICTION**

182 Birthmarks are just that, marks that are present when a baby is born. Usually they are small and flat and most people wouldn't notice them. Port wine stains are birthmarks made of tiny blood vessels. Usually they don't cause any trouble but they can be quite dark and some people can be very sensitive about having them. Port wine stains are called that because they are a dark red just like a dark red wine. They are permanent and cannot be treated.

✓ **FACT** **OR** **FICTION**

183

Ringworm is called ringworm because it forms a ring on the skin and is caused by a worm. It's a very contagious condition and is passed by direct skin contact with an infected person or animal. It starts as a little pimple and then the ring gets bigger. The outside edge of the ring will be red and the inside will clear up. It can be itchy and sore.

✓ **FACT**   **OR**   **FICTION**

184

Not every creature with eight legs is a spider. There is a little mite that has eight legs and causes a condition called scabies. The mites most like to live in the skin, but can live in clothes or sheets for a day or two if they have to. Scabies causes tiny blisters which can be itchy, and even more itchy after a hot bath or at night. Scratch. Scratch. Scratch.

✓ FACT   OR   FICTION

Freaky Fact or Fiction

185

Palaeontology (pale-ee-en-toll-o-gee) is the study of fossils. Pathology (path-oll-o-gee) is the study of disease. Palaeopathology (pale-ee-o-path-oll-o-gee) is the study of disease in fossils. Palaeopathologists study the fossils or mummified remains of ancient bodies to see if they can work out what killed them. They also look for things like arthritis and bone healing.

 ✓ **FACT**   **OR**   **FICTION**

186

Have you ever felt threatened? Or scared? Scared enough to 'run for your life'? Adrenalin (ah-dren-ah-lin) helps you prepare for either flight or fight. Adrenalin or epinephrine (epp-in-eff-rin) is known as the emergency hormone. It closes down the littlest blood vessels to make sure there is plenty of blood for the muscles to work. It also increases the heart rate so if you need to run, you can.

✓ **FACT** **OR** **FICTION**

Freaky Fact or Fiction

187

Use it or lose it! This saying reminds us that you need to keep moving, keep practising, keep whatevering if you want to stay on top of your game. If you stop training for your favourite sport, you won't be quite as good when you start again. In amblyopia (am-blee-oh-pee-uh) or lazy eye, your brain may favour the non-lazy eye and may eventually start to ignore the signals from the lazy one. That's why lazy eye is usually treated early, so the brain stops playing favourites!

✓ **FACT** **OR** **FICTION**

188 A majority of people write with their right hand. In the old days it was actually thought evil to write with your left hand. Some teachers tied children's hands behind their back so they HAD to use their right hand. Crazy! But some clever people can write with both hands. They might also be able to throw and catch equally well with both arms. They're called ambidextrous (am-bee-dex-truss). That's who you want on your team!

 FACT **OR** **FICTION**

189

How tidy is your room? Do all your clothes fit tidily into your drawers or cupboard? What happens to bits that stick out? They get squished when you shut the drawers or door. A similar thing can happen in your body. Many of your organs are held together in a big sack called the peritoneum (perri-to-nee-um). There are a few openings for bits to get in and out (otherwise you wouldn't be able to eat). Sometimes bits get squished near the openings. This is called a hernia.

✓ **FACT**   **OR**   **FICTION**

190

Itchy, itchy, itchy! But don't scratch! Most children are now vaccinated against illnesses like chickenpox. But some kids can still catch it. And it can spread like wildfire through families and classrooms. Mostly it's just a pesky thing with a short time of feeling unwell and some very, very itchy little blisters. But the virus that causes it is a sneaky thing. After you recover from chickenpox, the virus can hide near the spine for years. Then, when you least expect it, it can strike! When it does this it's called shingles. Shingles can cause blisters all over the body, but it doesn't hurt at all.

 ✓ **FACT** **OR** **FICTION**

Freaky Fact or Fiction

191

Do you get nervous when you have to do something new, or if you have to give a talk to the whole class or to the whole school? It's quite normal to feel nervous when something is new or different. Your heart beats faster, your tummy feels fluttery and your palms get sweaty. Believe it or not, it can actually be a good thing. It's your body preparing for the unknown. In this case, it's the fear that you will make an idiot of yourself in front of your friends. Often the anticipation of doing something new is much worse than actually doing it.

 ✓ **FACT** **OR** **FICTION**

192

The nervous system isn't the thing that makes you nervous. It's the name given to the whole network of nerves in your body. There's the brain and the spinal column. The spine is like an upside-down tree trunk with 31 pairs of nerves branching outwards. They branch, then branch again, into smaller and smaller nerves reaching all the way to your fingertips and to the tips of your toes. Amazing!

✓ **FACT** **OR** **FICTION**

193

torms are amazing to watch. All those dark clouds and flashes of lightning that light up the sky. Of course, it wouldn't be so much fun if you were out in the storm getting blown about and rain-soaked. It would be even less fun if you were at risk of being hit by lightning. But it can happen. People hit by lightning sometimes receive what are called entry and exit burns. They have burns where the lightning hits, and burns where the lightning exits to the ground, and nothing in between.

 ✓ **FACT** **OR** **FICTION**

194

Eat well and you'll grow up big and strong. Who hasn't been told that? Well, there is one time when you don't want to be the tallest. That's if you get caught in the middle of an open field during a wild storm. Lightning likes tall things. And if you're the tallest thing around... watch out! So, here's some advice if you are caught in a storm and can't find a safe place to shelter. Crouch down and make yourself as small as you can. Oh, one more thing. Boys are four times as likely as girls to be hit by lightning.

✓ **FACT** **OR** **FICTION**

195

Twenty20 is a form of cricket where each side has 20 overs. It is also the description of 'normal' vision. Someone with 20/20 vision can see the smallest letters on an eye chart from 20 ft away. It's also called 6/6 vision in some countries because the chart is 6 m away. The first number refers to what scientists have worked out is the standard for most people, and the second number refers to your results when tested. So if you could only see clearly at a distance of 10 ft, then your vision would be 20/10.

✓ **FACT**   **OR**   **FICTION**

196

Cats can see well at night. Other animals too – particularly those that are more active after the sun has gone down. Human vision is best when there is light, but most people adjust fairly quickly to darkness. Pupils dilate (get bigger) and we can begin to see things that seconds ago were invisible. But not everyone. Some people have night-blindness, where they can't see much at all in dim light or darkness. It's lucky we humans don't have to rely on our night-sight for finding food.

 FACT **OR** **FICTION**

Freaky Fact or Fiction

197 Have you ever been to a museum and seen organs floating in clear liquid? Scientists preserve things in a special form of alcohol so they can study them. Your brain and spine also float in a clear liquid called cerebrospinal (ser-e-bro-spy-nal) fluid (CSF). The fluid around your brain is a form of alcohol too. There's not that much of it, just over half a glass, but it's very important. If you bump your head, the CSF helps to absorb the shock and protect the brain from injury.

 ✓ FACT OR FICTION

198

Some people say the silliest things! Mostly it's because they forget to put their brain in gear before opening their mouth. But for people with Tourette syndrome things are different. They may have one 'tic' (muscle twitch), or they may have many. They may also say words unexpectedly. Sometimes they'll say the same word over and over, or swear. Sometimes they don't say words, but whistle or hiss. With practice and concentration it can be cured.

✓ FACT OR FICTION

199

Depending on where you live, the air temperature can vary wildly from very cold (below freezing) to very hot, sometimes in a single day. Your body doesn't like to change temperature; in fact, it really likes to stay the same (think about how horrible a fever feels). The normal body temperature is about 37°C (98.6°F). Temperatures of 40°C or 30°C will both cause hallucinations. Too hot causes violent fever. Too cold makes sufferers want to pull off all their clothes because their skin feels like it's burning.

 ✓ FACT OR FICTION

200

Your body has a language of its own. It can't actually say any words but it communicates anyway. Sit in the playground at school. Watch the students around you. You can tell if someone is happy or sad by the expression on their face, but also by the way they sit or walk. When you go home, practise in front of the mirror. Hold your head up high and pull your shoulders back. Not only will you be taller, you'll also look much more confident.

 ✓ FACT   **OR**   **FICTION**

201

Hepatitis is inflammation of the liver. People with hepatitis can sometimes have yellow skin. You can tell them apart from people who have bad fake tan by looking in their eyes. The whites of the eyes of people with hepatitis will also have a yellowish tinge. Doctors used to think there was only one type of hepatitis, but now we know there are three: Hepatitis A, Hepatitis B and Hepatitis C.

✓ **FACT**   **OR**   **FICTION**

202

Can you walk in a straight line? Of course you can, particularly if you have a straight line to walk along. If you are right-handed, your right leg is probably a bit stronger than your left. If you are left-handed, your left leg will be stronger. When you walk, your strong leg takes slightly longer steps than your weaker one. So, if you were lost in the desert, you'd probably end up walking in a circle.

 FACT **OR** **FICTION**

203

You've made it to the green of the first hole at the golf course. You put away your iron and get out your putter. You work out the exact angle you need to hit the ball so it goes into the hole. Then suddenly, you get the yips and your ball goes the wrong way. The yips? The yips are involuntary muscle spasms that wreck your shot. Scientists are unsure what causes them.

✓ **FACT** OR **FICTION**

204

Palindromes (pal-in-dromes) are words or phrases that are the same whether they're read from left or right. For example, kayak, racecar, or the phrase 'never odd or even'. Palindromic rheumatism (pal-in-droh-mic roo-ma-tizm) causes sudden inflammation in and around joints which comes and goes. It finishes the way it begins – suddenly. That's why it's called palindromic.

✓ **FACT** **OR** **FICTION**

Answers

1. Fact.

2. Fact.

3. Fact.

4. Fact.

5. Fact.

6. Fact.

7. Fact.

8. **Fiction.** Fingernails, toenails and hair only appear to grow longer after death because the skin dries and shrinks away from them.

9. **Fiction.** Your sense of smell also helps you taste food, and the average number of tastebuds on your tongue is between 2000 and 8000.

10. **Fiction.** Your left lung is smaller than your right lung so there is room for your heart. Your left lung has only two sections: the superior lobe and the inferior lobe.

11. Fact.

12. **Fiction.** There are more than 14 billion nerve cells in the outer layer of the cerebrum.

13. Fact.

14. Fact.

15. **Fiction.** The epiglottis is at the top of your breathing tube (trachea).

16. Fact.

17. Fact.

18. Fact.

19. Fact.

20. Fact.

21. **Fiction.** There are actually between two million and five million sweat glands on your skin.

22. Fact.

23. Fact.

24. Fact.

25. **Fiction.** Mostly these muscles are too small and weak to move your ears at all. Very few people can move their ears.

26. **Fiction.** The cartilage and skin that make up nose and ears do become saggier with age and this may make them look bigger.

27. Fact.

28. Fact.

29. Fiction. The soft spots can be there until you are about two years old.

30. Fact.

31. Fact.

32. Fiction. Straight hair appears round, and curly hair looks almost flat, like a ribbon.

33. Fact.

34. Fact.

35. Fact.

36. Fact.

37. Fact.

38. Fiction. Each glass of milk can lead to two to four cups of hydrogen gas in the belly.

39. Fact.

40. Fact.

41. Fact.

42. Fact.

43. Fact.

44. Fiction. There are two record-holders for loud burps. Jodie Parks (USA) and Paul Hunn (UK) can both burp louder than a pneumatic drill. Paul Hunn's burp was a little bit louder at 107.1 decibels.

45. Fact.

46. Fact.

47. Fact.

48. Fact.

49. Fact.

50. Fact.

51. Fact.

52. Fiction. There are 600 muscles in your body and about 40 per cent of the body's weight is muscle.

53. Fact.

54. Fact.

55. Fact.

56. Fact.

57. Fact.

58. Fact.

59. Fact.

60. Fiction. These flatworms lay between 300 and 3500 eggs per day.

61. Fact.

62. Fact.

63. Fact.

64. Fact.

65. Fiction. Humans need exposure to sunlight to help their bodies produce Vitamin D, which helps to grow strong bones.

66. Fact.

67. Fact.

68. Fact.

69. Fact.

70. Fact.

71. Fact.

72. Fiction. Freckles usually don't appear until about five years of age.

73. Fiction. Your nose runs because the tear ducts, which drain fluid from the eyes, open directly into your nose.

74. Fact.

75. Fiction. In general, the larger an animal, the slower its heart rate. A canary's heart beats at about 1000 beats per minute and an elephant's heart beats at around 25 beats per minute.

76. Fact.

77. Fact.

78. Fact.

79. Fact.

80. Fact.

81. Fact.

82. Fact.

83. Fact.

84. Fact.

85. Fact.

86. Fact.

87. Fiction. Onychophagia literally means nail-eating.

88. Fiction. Albinism can impair vision and make sufferers more sensitive to light.

89. Fact.

90. Fact.

91. Fact.

92. Fiction. Boys are 20 times more likely to be red-green colour-blind.

93. Fact.

94. Fact.

95. Fact.

96. Fact.

97. Fiction. At 150km/h (93 mi/h) your sneeze would be Category 2: destructive. Likely to cause significant damage to trees and caravans or travel trailers.

98. Fact.

99. Fact.

100. Fact.

101. Fact.

102. Fiction. The sphincter in the eye does tighten in bright light, but there is also a dilator muscle that works hard to open the pupil when there is less light.

103. Fiction. They won't be able to remember what you said, because they didn't hear it. They were asleep, remember!

104. Fact.

105. Fiction. The condition is called cataplexy.

106. Fact.

107. Fact.

108. Fact.

109. Fact.

Answers

● ● ● ● ● ● ● ● ● ● ● ● ● ● ● ● ● ● ●

110. Fact.

111. Fiction. A hair ball might sit in the stomach for years.

112. Fiction. It's called a butterfly dressing because that's what it looks like.

113. Fact.

114. Fact.

115. Fact.

116. Fact.

117. Fact.

118. Fiction. There is an adult form of rickets called osteomalacia (oss-tee-oh-mal-ay-sha).

119. Fiction. Tonsils are only removed when someone repeatedly suffers severe tonsillitis and it makes them very sick.

120. Fact.

121. Fiction. Tonsils can't be replaced.

122. Fiction. The core of a long bone is spongy and is full of little holes.

123. Fact.

124. Fiction. There is one exception. The pulmonary artery carries darker (less oxygen) blood to the lungs. The pulmonary veins carry the brighter (more oxygen) blood back to the heart.

125. Fact.

126. Fact.

127. Fact.

128. Fact.

129. Fact.

130. Fact.

131. Fact.

132. Fiction. Fever is actually good. It's your body's way of mobilising all the virus- and infection-fighting cells to make you better.

133. Fact.

134. Fact.

135. Fact.

136. Fact.

137. Fact.

138. Fact.

139. Fact.

140. Fiction. Humans can sense more than 400 smells.

141. Fact.

142. Fact.

143. Fact.

144. Fiction. New studies suggest that the ability to roll your tongue may involve more than just genetics (i.e. inheriting the ability) – your parents might carry a 'tongue-rolling' gene!

145. Fact.

146. Fact.

147. Fiction. It's called sinus squeeze, because it feels like someone is squeezing the front of your head.

148. Fact.

149. Fact.

150. Fact.

151. Fact.

152. Fiction. The uvula is thought to help block the back of the nose when you swallow and also to have some role in singing.

153. Fact.

154. Fact.

155. Fact.

156. Fact.

157. Fact.

158. Fact.

159. Fiction. The globe of a fully grown eye weighs about 7.5 g (0.25 oz), or the same as 1½ teaspoons of sugar.

Answers

● ●

160. Fact.

161. Fiction. Rod cells (130 million each eye) tell you about shape and size etc. Cone cells (seven million) tell you about colour and detail.

162. Fact.

163. Fact.

164. Fact.

165. Fact.

166. Fact.

167. Fact.

168. Fact.

169. Fact.

170. Fiction. Children with progeria can start losing hair by the time they have their FIRST birthday.

171. Fact.

172. Fact.

173. Fiction. The study of fingerprinting is called 'dactyloscopy' (dak-till-oh-scoh-pee). A dactylologist uses their fingers and hands to 'speak' to hearing impaired people (sign language).

174. Fact.

175. Fact.

176. Fact.

177. Fact.

178. Fact.

179. Fact.

180. Fiction. When the cuff is first loosened, the blood will sound gurgly like a little waterfall. The second measurement is taken when the blood sound is back to normal.

181. Fact.

182. Fiction. Port wine stains can be treated with lasers.

183. Fiction. Ringworm is caused by a fungus, not a worm.

184. Fact.

185. Fact.

186. Fact.

187. Fact.

188. Fact.

189. Fact.

190. Fiction. Shingles blisters appear in a horizontal line and can be very painful for months.

191. Fact.

192. Fact.

193. Fact.

194. Fact.

195. Fiction. The first number is about YOUR sight and the second is about what's 'normal' for most people. So if you could only see clearly at a distance of 10 ft, then your vision would be 10/20.

196. Fact.

197. Fiction. CSF is mainly water.

198. Fiction. Tourette syndrome cannot be cured, but concentration and medication can help ease the symptoms.

199. Fact.

200. Fact.

201. Fiction. There is also Hepatitis D, E, F and G.

202. Fiction. Even in the desert, the land isn't completely flat, so if you walk in a circle, it's not because one leg is slightly stronger than the other. One study even suggests that if you're lost you won't be able to walk in a straight line without the sun or the moon as a reference point.

203. Fact.

204. Fiction. Palindromic in this sense means 'happening again' or 'getting worse'.

Sources

1. Gray's Anatomy Online, www.bartleby.com/107/, 2000

2. Encyclopaedia Britannica Online, www.britannica.com, 2010

3. Encyclopaedia Britannica Online, 2010

4. Encyclopaedia Britannica Online, 2010

5. Encyclopaedia Britannica Online, 2010

6. Gray's Anatomy Online, 2000

7. Encyclopaedia Britannica Online, 2010; Gray's Anatomy Online, 2000

8. Encyclopaedia Britannica Online, 2010

9. Encyclopaedia Britannica Online, 2010

10. Gray's Anatomy Online, 2000

11. Encyclopaedia Britannica Online, 2010

12. Encyclopaedia Britannica Online, 2010

13. Encyclopaedia Britannica Online, 2010

14. Encyclopaedia Britannica Online, 2010

15. Encyclopaedia Britannica Online, 2010

16. Encyclopaedia Britannica Online, 2010

17. Encyclopaedia Britannica Online, 2010

18. Encyclopaedia Britannica Online, 2010

19. Encyclopaedia Britannica Online, 2010

20. Gray's Anatomy Online, 2000

21. Encyclopaedia Britannica Online, 2010; Gray's Anatomy Online, 2000

22. Encyclopaedia Britannica Online, 2010

23. Encyclopaedia Britannica Online, 2010

24. Encyclopaedia Britannica Online, 2010; Gray's Anatomy Online, 2000

25. Gray's Anatomy Online, 2000

26. Encyclopaedia Britannica Online, 2010

27. Encyclopaedia Britannica Online, 2010; Gray's Anatomy Online, 2000

28. Gray's Anatomy Online, 2000

29. Encyclopaedia Britannica Online, 2010

30. Encyclopaedia Britannica Online, 2010

31. Encyclopaedia Britannica Online, 2010

32. Encyclopaedia Britannica Online, 2010

33. Encyclopaedia Britannica Online, 2010

34. Encyclopaedia Britannica Online, 2010

35. Encyclopaedia Britannica Online, 2010

36. Encyclopaedia Britannica Online, 2010

37. Encyclopaedia Britannica Online, 2010

38. Encyclopaedia Britannica Online, 2010

39. Gray's Anatomy Online, 2000

40. 'Interesting Facts', Australian Podiatry Association (Vic), www.podiatryvic.com.au, 2010

41. *Guinness World Records 2009* (book), 2009

42. *Guinness World Records 2009* (book), 2009; Guinness World Records, www.guinnessworldrecords.com

43. *Guinness World Records 2010* (book), 2010; Guinness World Records, www. guinnessworldrecords.com

44. *Guinness World Records 2009* (book), 2009

45. *Guinness World Records 2009* (book), 2009

46. *Guinness World Records 2009* (book), 2009; Guinness World Records, www. guinnessworldrecords.com

47. *Guinness World Records 2009* (book), 2009

48. *Guinness World Records 2009* (book), 2009

49. *Guinness World Records 2009* (book), 2009

50. *Guinness World Records 2009* (book), 2009

51. *Guinness Book of Knowledge* (book), 1997

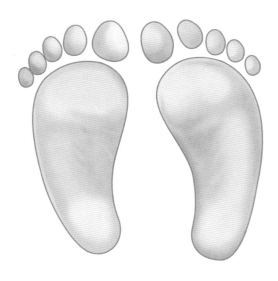

Sources

52. Encyclopaedia Britannica Online, 2010

53. Encyclopaedia Britannica Online, 2010

54. Encyclopaedia Britannica Online, 2010

55. Encyclopaedia Britannica Online, 2010

56. Encyclopaedia Britannica Online, 2010

57. Encyclopaedia Britannica Online, 2010

58. Encyclopaedia Britannica Online, 2010

59. Encyclopaedia Britannica Online, 2010

60. Encyclopaedia Britannica Online, 2010

61. Merck & Co, Inc, www.merck.com, 2010

62. Mayo Clinic.com, www.mayoclinic.com, 2010

63. Merck & Co, Inc, www.merck.com, 2010

64. Merck & Co, Inc, www.merck.com, 2010; Healthy Children, www.healthychildren.org, 2010

65. Merck & Co, Inc, www.merck.com, 2010

66. Merck & Co, Inc, www.merck.com, 2010

67. Mayo Clinic.com, www.mayoclinic.com, 2010

68. 'Meibomian Gland Dysfunction', Cornea & Contact Lens Society of New Zealand, www.contactlens.org.nz, 2010

69. Gray's Anatomy Online, 2000

70. Encyclopaedia Britannica Online, 2010

71. Mayo Clinic.com, www.mayoclinic.com, 2010

72. Encyclopaedia Britannica Online, 2010

73. Encyclopaedia Britannica Online, 2010

74. Encyclopaedia Britannica Online, 2010

75. Encyclopaedia Britannica Online, 2010

76. Zaid A Wani, Abdul W Khan, Aijaz A Baba, Hayat A Khan, Qurat-ul Ain Wani and Rayeesa Taploo, 'Cotard's Syndrome and delayed diagnosis in Kashmir, India', *International Journal of Mental Health Systems* (online journal), vol. 2 (1), 2008

77. 'Trimethylaminuria: A Case Report', *Dermatology* (online magazine), January 2007

78. Encyclopaedia Britannica Online, 2010

79. Mayo Clinic.com, www.mayoclinic.com, 2010

80. Encyclopaedia Britannica Online, 2010

81. 'Anthropometry & Biometrics', NASA, http://msis.jsc.nasa.gov, 2008

82. Encyclopaedia Britannica Online, 2010

83. University of Maryland Medical Center, www.umm.edu/conjoined_twins/facts.htm, 2011

84. 'Managing House Dust Mites', University of Nebraska-Lincoln, www.unl.edu, 2010

85. 'Managing House Dust Mites', University of Nebraska-Lincoln, www.unl.edu, 2010

86. Merck & Co, Inc, www.merck.com, 2010

87. *Blakiston's Medical Dictionary*, 1979

88. Merck & Co, Inc, www.merck.com, 2010; Encyclopaedia Britannica Online, 2010

89. *Catalyst*, ABC TV, 15 April 2010

90. Merck & Co, Inc, www.merck.com, 2010

91. Encyclopaedia Britannica Online, 2010

92. Encyclopaedia Britannica Online, 2010

93. Gray's Anatomy Online, 2000

94. Encyclopaedia Britannica Online, 2010

95. 'Interesting Facts', Canadian Shark Research Laboratory, www.marinebiodiversity.ca/shark, 2010

96. 'Sneezing', Australasian Society of Clinical Immunology & Allergy, www.allergy.org.au, 2010

Sources

97. 'Sneezing', Australasian Society of Clinical Immunology & Allergy, www.allergy.org.au, 2010

98. Encyclopaedia Britannica Online, 2010; Merck & Co, Inc, www.merck.com, 2010

99. Encyclopaedia Britannica Online, 2010

100. Encyclopaedia Britannica Online, 2010

101. Encyclopaedia Britannica Online, 2010

102. Encyclopaedia Britannica Online, 2010

103. Encyclopaedia Britannica Online, 2010

104. Encyclopaedia Britannica Online, 2010

105. Encyclopaedia Britannica Online, 2010

106. Encyclopaedia Britannica Online, 2010

107. Encyclopaedia Britannica Online, 2010

108. Merck & Co, Inc, www.merck.com, 2010

109. *Guinness Book of Knowledge* (book), 1997

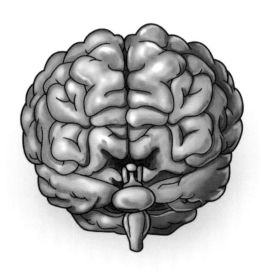

110. Mayo Clinic.com, www.mayoclinic.com, 2010

111. Mayo Clinic.com, www.mayoclinic.com, 2010

112. Mayo Clinic.com, www.mayoclinic.com, 2010

113. Encyclopaedia Britannica Online, 2010

114. Encyclopaedia Britannica Online, 2010

115. Encyclopaedia Britannica Online, 2010

116. Encyclopaedia Britannica Online, 2010

117. Mayo Clinic.com, www.mayoclinic.com, 2010

118. Encyclopaedia Britannica Online, 2010

119. Mayo Clinic.com, www.mayoclinic.com, 2010

120. Encyclopaedia Britannica Online, 2010

121. Linda Calabresi, *Insiders Human Body* (book), 2007

122. Linda Calabresi, *Insiders Human Body* (book), 2007

123. Linda Calabresi, *Insiders Human Body* (book), 2007

124. Linda Calabresi, *Insiders Human Body* (book), 2007

125. Linda Calabresi, *Insiders Human Body* (book), 2007

126. Linda Calabresi, *Insiders Human Body* (book), 2007

127. Encyclopaedia Britannica Online, 2010

128. Encyclopaedia Britannica Online, 2010

129. Encyclopaedia Britannica Online, 2010

130. Encyclopaedia Britannica Online, 2010

131. 'Bone Fracture Healing Explained', Physio Room.com, www.physioroom.com, 2010

132. Encyclopaedia Britannica Online, 2010

133. Encyclopaedia Britannica Online, 2010

134. Encyclopaedia Britannica Online, 2010

135. Encyclopaedia Britannica Online, 2010

136. Encyclopaedia Britannica Online, 2010

137. Encyclopaedia Britannica Online, 2010

138. Encyclopaedia Britannica Online, 2010

139. Encyclopaedia Britannica Online, 2010

140. Encyclopaedia Britannica Online, 2010

141. Encyclopaedia Britannica Online, 2010

142. Merck & Co, Inc, www.merck.com, 2010

143. 'Foot Odour', Australian Podiatry Association (Vic), www.podiatryvic.com.au, 2010

144. 'Ask a Geneticist', The Tech Museum, www.thetech.org/genetics, 2005

145. Encyclopaedia Britannica Online, 2010

146. Encyclopaedia Britannica Online, 2010

147. Encyclopaedia Britannica Online, 2010

148. Encyclopaedia Britannica Online, 2010

149. Encyclopaedia Britannica Online, 2010

Sources

150. Encyclopaedia Britannica Online, 2010

151. Encyclopaedia Britannica Online, 2010

152. Encyclopaedia Britannica Online, 2010

153. *Collins English Dictionary Edition 3*, 1995

154. Encyclopaedia Britannica Online, 2010

155. Encyclopaedia Britannica Online, 2010

156. Encyclopaedia Britannica Online, 2010

157. Encyclopaedia Britannica Online, 2010; Doctors Corner, http://your-doctor.com, 2005

158. *Macquarie Concise Dictionary Fourth Edition*, 2006; Doctors Corner, http://your-doctor.com, 2005

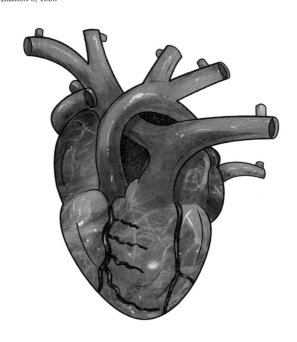

159. Encyclopaedia Britannica Online, 2010

160. Encyclopaedia Britannica Online, 2010

161. Encyclopaedia Britannica Online, 2010

162. Encyclopaedia Britannica Online, 2010

163. 'Making a new artificial eye', Artificial Eyes and Ocularists, http://geelen.com.au, 2010

164. *Collins English Dictionary Edition 3*, 1995

165. *Guinness World Records 2009* (book), 2009; Elaine Davidson, www.elainedavidson.co.uk, 2010

166. *Guinness World Records 2009* (book), 2009

167. Encyclopaedia Britannica Online, 2010

168. Encyclopaedia Britannica Online, 2010

169. Guinness World Records, www.guinnessworldrecords.com, 2010

170. Encyclopaedia Britannica Online, 2010

171. Mayo Clinic.com, www.mayoclinic.com, 2010

172. Gray's Anatomy Online, 2000

173. *Collins English Dictionary Edition 3*, 1995

174. Encyclopaedia Britannica Online, 2010

175. www.oxfordreference.com, 2010

176. The Merck Manual Medical Library, www.merck.com/mmpe/index.html, 2010

177. www.oxfordreference.com, 2010

178. Encyclopaedia Britannica Online, 2010

179. www.oxfordreference.com, 2010

180. www.oxfordreference.com, 2010

181. Royal Children's Hospital Melbourne, www.rch.org.au/kidsinfo, 2010

182. Royal Children's Hospital Melbourne, www.rch.org.au/kidsinfo, 2010

183. Royal Children's Hospital Melbourne, www.rch.org.au/kidsinfo, 2010

184. Royal Children's Hospital Melbourne, www.rch.org.au/kidsinfo, 2010

185. www.oxfordreference.com, 2010

186. Encyclopaedia Britannica Online, 2010

187. Mayo Clinic.com, www.mayoclinic.com, 2010

188. Encyclopaedia Britannica Online, 2010

189. Mayo Clinic.com, www.mayoclinic.com, 2010

190. Mayo Clinic.com, www.mayoclinic.com, 2010

191. Encyclopaedia Britannica Online, 2010

192. Encyclopaedia Britannica Online, 2010

193. 'Human Voltage – what happens when people and lightning converge', NASA Science, http://science.nasa.gov, 2010

194. 'Human Voltage – what happens when people and lightning converge', NASA Science, http://science.nasa.gov, 2010

Sources

195. Mayo Clinic.com, www.mayoclinic.com, 2010

196. Encyclopaedia Britannica Online, 2010

197. Encyclopaedia Britannica Online, 2010

198. Encyclopaedia Britannica Online, 2010

199. Nick Arnold and Tony de Saulles, *The Horrible Science of You* (book), 2009

200. Nick Arnold and Tony de Saulles, *The Horrible Science of You* (book), 2009

201. Encyclopaedia Britannica Online, 2010

202. Mayo Clinic.com, www.mayoclinic.com, 2010; 'Lost people really do walk in circles: study', Reuters, www.reuters.com, 2009

203. Mayo Clinic.com, www.mayoclinic.com, 2010

204. Mayo Clinic.com, www.mayoclinic.com, 2010

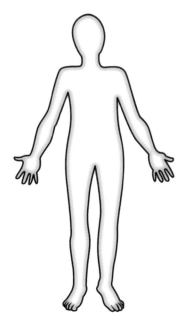